T0133297

A Distributed Virtual Reality System for Spatial Updating

Concepts, Implementation, and Experiments

Dissertation

zur Erlangung des akademischen Grades

Doktor der Naturwissenschaften

der Technischen Fakultät der Universität Bielefeld
vorgelegt von Markus von der Heyde am 19. September 2000

Advisers:
Prof. Heinrich H. Bülthoff[1]
Prof. Gerhard Sagerer[2]
Prof. Dana Ballard[3]

[1]Max-Planck-Institute for Biological Cybernetics, Tübingen
[2]Faculty of Computer Science and Applied Natural Sciences, University of Bielefeld
[3]Computer Scince Department, University of Rochester, NY

Die Deutsche Bibliothek – CIP-Einheitsaufnahme

Heyde, Markus von der:
A distributed virtual reality system for spatial updating : concepts, implementation, and experiments / vorgelegt von Markus von der Heyde. - Berlin : Logos-Verl., 2001

(MPI Series in biological cybernetics ; Bd. 2)
Zugl.: Bielefeld, Univ., Diss., 2000
ISBN 3-89722-781-9

ISBN 3-89722-781-9
ISSN 1618-3037

Logos Verlag Berlin
Comeniushof, Gubener Str. 47,
10243 Berlin
Tel.: +49 030 42 85 10 90
Fax: +49 030 42 85 10 92
INTERNET: http://www.logos-verlag.de

Abstract

A Distributed Virtual Reality System for Spatial Updating
Concepts, Implementation, and Experiments

PhD Thesis of Markus von der Heyde

Introduction: Over the course of evolution humans as well as other animals learned to navigate through complex environments. Such navigation had two main goals: to find food and to find the way back to shelter. For most moving organisms it is important to know their location in the world and maintain some internal representation of it. For higher species it is most likely that multiple sensory systems provide information to solve this task. Consequently, to study human behavior in a complex environment it is important that the experimenter has full control over the stimulus for multiple senses. Furthermore, it is crucial to guarantee the following: A) The stimulus, and the information it conveys has to be precisely controllable; B) The experimental conditions have to be repeatable; and C) The stimulus conditions have to be independent of the individual characteristics of the observer.
Virtual Environments have to some degree offered a solution for these demands. Recently, it has become increasingly possible to conduct psychophysical experiments with more than one sensory modality at a time. In this thesis, Virtual Reality (VR) technology was used to design multi-sensory experiments which look into some aspects of the complex multi-modal interactions of human behavior.

Contents: The first part of this PhD thesis describes a Virtual Reality laboratory which was built to allow the experimenter to stimulate four senses at the same time: vision, acoustics, touch, and the vestibular sense of the inner ear. Special purpose equipment is controlled by individual computers to guarantee optimal performance of the modality specific simulations. These computers are connected in a network functioning as a distributed system using asynchronous data communication. The second part of the thesis presents two experiments which investigate the ability of humans to perform spatial updating. These experiments contribute new scientific results to the field and serve, in addition, as proof of concept for the VR-lab. More specifically, the experiments focus on the following main questions: A) Which information do humans use to orient in the environment and maintain an internal representation about the current location in space?; B) Do the different senses code their percept in a single spatial representation which is used across modalities, or is the representation modality specific?

Results and Conclusions: The experimental results allow the following conclusions: A) Even without vision or acoustics, humans can verbally judge the distance traveled, peak velocity, and to some degree even maximum acceleration using relative scales. Therefore, they can maintain a good spatial orientation based on proprioception and vestibular signals; B) Learning the sequence of orientation changes with multiple modalities (vision, proprioception and vestibular input) enables humans to reconstruct their heading changes from memory. In situations with conflicting cues, the maximum percept from either of the modalities had a major influence on the reconstruction. Most of the naïve subjects did not notice any conflicts between modalities. In total, this seems to suggest that there is a single spatial reference frame used for spatial memory. One possible model for cue integration might be based on a dynamically weighted sum of all modalities which is used to come up with a coherent percept and memory for spatial location and orientation.

Contents

Chapter 1

Introduction

The main question addressed in this thesis is: **"How do we know where we are?"**. Normally, humans know where they are with respect to their individual environment. The overall perception of this environment results from the integration of multiple sensory modalities. Here we use Virtual Reality to study the interaction of several senses and explore the way these senses might be integrated into a coherent perception of spatial orientation and location. This thesis will describe a Virtual Reality laboratory, its technical implementation as a distributed network of computers, and several basic experiments designed to investigate questions of spatial orientation.

This introductory chapter is divided into four main parts. The first three of them will focus on definitions, basic terminology, and examples for Virtual Reality, Distributed Systems, and Psychophysics. The last section will give a brief summary of the remaining chapters of this thesis. It introduces the Motion-Lab, the specific experimental questions and the main results.

1.1 Virtual Reality and Virtual Environments

The Latin etymological root for the word *virtual* means "*existing by power and possibility, able to work or cause, apparently or seemingly*". How can this description be combined with the term *reality* which is the existence of all facts around us? Researchers have defined Virtual Reality, for example, as:

> "*... a high-end user interface that involves real-time simulation and interactions through multiple sensorial channels. These sensorial modalities are visual, auditory, tactile, smell, taste, etc.*" (Burdea, 1993)

In addition to the quote given above, Burdea and Coiffet (1994) summarized attempts at defining VR and stated clearly what Virtual Reality is **not**. The authors rejected definitions where the sense of "presence", the immersion at a remote location, is dominating, as well as definitions where parts of the real environment are replaced or enhanced by simulated features ("enhanced reality"). In addition, all statements associating the definition of Virtual Reality with a specific set of interaction devices like head mounted displays, position sensing gloves or joysticks are not adequate, since those tools can easily be exchanged or used in applications not at all connected to Virtual Reality.

1.1.1 Definitions and terminology

Today, following the definition given above, the functionality of **Virtual Reality (VR)** is often described as complex computer simulation which exchanges the natural reality of a potential user with a simulated version. The exchange is limited, in most cases, to some, but not all of the senses the user could experience in the simulation. Visual simulations are typically the main part of today's Virtual Reality applications. The visual simulation tries to mimic the relevant aspects of the visual world, creating a simplified version. Natural looking sceneries (Virtual Environments) are the goal of many research projects, a goal that has not so far been achieved due to the overwhelming complexity of even a simple outdoor scene. Nonetheless, the existing visual simulations cover some important features of the visual world like reflections, shading, shadows, and natural scene dynamics. However, Virtual Reality is not confined to visual simulations, but also must include the user, allowing active behavior inside the simulation. The **user** interacts with the simulation via specialized interfaces. Actions of the user cause changes in the simulation and feedback is provided to let the user "immerse into another world". The sense of **presence** can be strengthened by real-time feedback involving the user in Virtual Reality. Very closely related to Virtual Reality is the term **Virtual Environment (VE)** which refers to the simulated environment itself. Virtual Environments can be presented with Virtual Reality technology in multiple sensory modalities.

In addition to the given definition of VR, Burdea and Coiffet (1994) summarized the history of VR. Already in the 1960's, the Sensorama (Heilig, 1960) provided color and 3D video, stereo sound, aromas, wind and vibrations in a motorcycle ride through New York (see Fig. 1.1). This device delivered full sensation in multiple modalities, but the user could not interact. Nonetheless, this historical device can be seen as be beginning of Virtual Reality. It provided a sensational experience in advance of many of today's systems.

Today's Virtual Reality systems actually got away from the presentation of odor and have concentrated on replacing the video input from the Sensorama with a simulated version which is capable of reacting to user input. The user in a Virtual Environment, or in Virtual Reality in general, can interact with the simulated world. The interaction is made possible by different devices for each sense.

VR interfaces have also changed since the 1960's, becoming smaller, more powerful, and lighter, thanks to the development of micro-technology. Visual simulations are presented mostly by projecting the simulated picture onto a screen in front of the user. Another approach is to make displays very small and integrate them into a helmet, letting the user see through a specialized optic system. These systems are called head mounted displays (HMD) an idea that dates back to Heilig (1960). Other equipment is used to enable users to interact with the simulated world. These interactive interfaces enable users to sense multiple sensory modalities simulated coherently by the computer. For example, virtual touch can be simulated by a device called PHANToM, which can be used for psychophysical experiments (e.g., von der Heyde and Häger-Ross, 1998). With this device virtual objects with a wide variety of properties can be touched by the user with the fingertip or a stylus. Other interfaces like virtual bikes or position sensing gloves are used to let the user navigate through virtual environments. The particular interfaces used in any VR setup mostly depend on the application and its goals.

Today's video games and multimedia applications very often use sound, videos, and 3D animations. Some interaction device like a computer mouse or joystick is typically used

Figure 1.1: The Sensorama provided acoustics, visual cues, and aroma for the user. Wind was applied with small fans and the seat could vibrate. It was developed and built by Heilig (1960) as a prototype to provide full sensory experience for one person. (The picture is taken from Burdea and Coiffet (1994).)

for control. Even though such applications use multiple modalities which can induce a sense of presence, we can still distinguish between them and true VR. One way of doing so is by comparing the degree of interaction (see Fig 1.2). For instance, we can switch a video recording on and can stop it any time, but there is no real influence on the picture we see. However, the picture will look realistic, since it is normally taken from the real world and not rendered synthetically. Moving further towards multimedia applications, the degree of interaction increases. In multimedia applications one can choose what to do next and where to look for new information. Nonetheless, the information presented as text, video or sound does not react to our input. In today's video games the degree of interaction is quite high. In a simulated 3D environment, the player can change his/her own position, collect objects, fight, and run. Some of the games are close to our definition of VR. Though in these games one can not feel objects the simulated ego picks up, but one can turn the objects around and use them as tools just as in the real world. The boundaries between VR, complex 3D animated games, and other similar applications become more and more vague and defined by the purpose of the application, and some might say by the costs of the system. Imagine a car race simulating a complex 3D environment, providing realistic sounds and using a force feedback steering wheel as input device. This game already simulates three modalities with immediate feedback to the user's reactions. Do other so called VR applications involve that much realism?

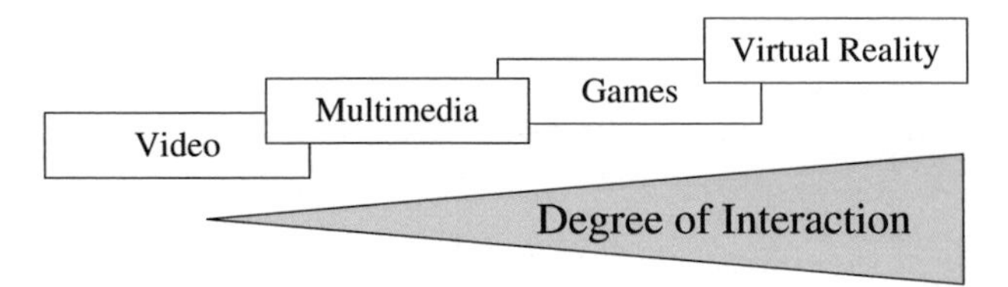

Figure 1.2: The degree of interaction can be used to distinguish between standard video, multimedia applications, action games, and today's Virtual reality applications.

1.1.2 General Applications

Virtual Reality can be used in many different fields with a very broad range of applications. Since the simulation depends only on the purpose of the application, nearly everything can be simulated, making use of the flexibility of the technology. The following overview can not cover all possibilities, but is intended to give a general impression of the range of possibilities.

Today's VR technology is used in **medicine** for surgical planing. Surgery in the brain, for example, is simulated beforehand, enabling the doctor to judge which brain areas should not be touched due to main blood vessels or important functions of those regions. Using VR technology in combination with nano-technology allows the surgeon to operate without direct access to the operation site. The surgeon can manipulate large scale devices and interact with VR interfaces. His actions are then transformed into actual movements of micro manipulators in the body. The detailed and realistic visualization makes both the planing, as well as the surgery, possible.

Other applications make use of VR for **training**. Due to the easy implementation of new scenarios, trainees can use multiple realistic models and learn a variety of situations. The technology reduces costs once the setup is installed and in use. The costs of investment for a driving simulator, for example, is much higher than for the actual car. However, the costs for maintenance and use are much lower in the long run. Especially dangerous situations are difficult to train with real vehicles, but can be easily accommodated in VR. In flight simulators, for example, researchers are interested in flight illusions which occur in situations when some visual cues are absent in real flights. The VR setup makes it possible to determine which situation will be perceived incorrectly without taking the risk of an actual crash. Many training applications, especially in the military sector, are forcing the development of high end application interfaces. The demand to make the situation as realistic as possible is driving VR technology forward.

The **visualization** of virtual scenery or data in general is not only important for military applications. To enable distant control in medicine and military applications it is necessary to provide high speed communication between the site of action and the distant interface. Alternatively, the visualization of a simulated version of the distant place temporarily replaces the real, exact data in order to provide sufficient feedback for the control. In other applications, natural resources, for example, are rendered in three dimensions to let engineers judge how to exploit them most efficiently. Oil and water resources are displayed taking environmental changes into account. Other projects generate virtual models of 3D surfaces from data collected by satellites. Those data can be integrated with weather models making the weather forecast more precise and understandable.

Visualization is also used for interactively displaying the **design** of new products. Rapid prototyping allows engineers to directly create real models after the design stages have been done in VR. New buildings are displayed in VR, guiding the user through a realistic environment. This allows the user to easily change colors and materials to get a good impression of how a room, or even a complete building, could look like in the future. Town development and planing incorporate high-end devices to design whole new parts of modern cities. The buildings are integrated into models of the existing town environment to enable a smooth integration of old and new housing.

In the future VR could be used in the **social sciences** to study human interaction (Loomis, Blascovich, and Beall, 1999). Psychologists use avatars (simulated persons) which can display natural looking behavior and facial expressions. Using this technology to study social interaction with the simulated person might give us a better understanding of developmental and psychological disorders. Changes of perception during human development can also be studied in VR. Clumsy children could be monitored before, during, and after a specialized training using VR setups to judge whether the training program is influencing their mobility. Similar studies have been successful using special computer games (Eliasson, Rösblad, and Häger-Ross, 2000).

It has also been demonstrated that VR can be used for **psychophysics** (see section 1.3). Properties of objects can be manipulated in a way that would be impossible or at least very expensive in a real environment. Stimuli for experiments can be independently controlled for all kinds of properties, allowing one to experimentally disentangle the individual influence of those cues on the human perception.

1.1.3 Examples of VR Labs

The following section describes six labs and their research agenda, which provides examples of VR systems. These labs were selected because the author either visited them or worked there for some time (Tübingen and Rochester). The web addresses of the labs are provided in appendix D.4.

VE Lab – Tübingen

At the MPI for Biological Cybernetics in Tübingen, there are several VR related projects. Of these projects, most are focusing on psychophysical experiments (Bülthoff, Foese-Mallot, and Mallot, 1997; Bülthoff and van Veen, 1999). One of the most impressive setups includes a cylindrical screen with a 7 m diameter. It is used for the projection of three pipes of the Onyx^2 Infinite Reality System. This mainframe computer has ten processors and 2.5 GB main memory. The three independent graphic pipelines allow the rendering of high fidelity virtual scenes. The three projections are blended by small overlapping zones with reduced light intensity. A steering wheel or a virtual bike can be used as the interaction device. More technical details can be found in Distler, van Veen, Braun, and Bülthoff (1997), van Veen, Distler, Braun, and Bülthoff (1998). Recent projects in this lab have been concerned with a simulation of the inner city of Tübingen (Virtual Tübingen). Spatial cognition is compared between the virtual version and the real city environment (Sellen, 1998). A different series of projects studied space perception in a virtual maze based on a hexagonal street configuration (Gillner and Mallot, 1996; Geiger, Gillner, and Mallot, 1997; Steck, 2000; Steck and Mallot, 2000). One strategy while navigating is path integration which was extensively studied in triangle completion experiments (Riecke, 1998; Riecke, van Veen, and Bülthoff, 1999; Bülthoff, Riecke, and van Veen, 2000; Riecke, van Veen, and Bülthoff, 2000). Other projects have studied driving abilities (Chatziastros, Wallis, and Bülthoff, 1997; Cunningham, von der Heyde, and Bülthoff, 2000a). New projects are including EEG and physiological parameters in studies of anxiety similar to those in Mühlberger, Herrmann, Pauli, Wiedemann, and Ellgring (1999). Psychophysical studies of velocity judgements are summarized by Distler (2000). All of the above projects use C or C++ code programming the VR application for the Performer or Vega visualization library. Most of the studies use models created with Multigen or Medit 3D modelling software.

VR Lab – Bielefeld

The VR Lab at the University of Bielefeld is working on natural human-machine interaction interfaces. Several projects centered around the proposal "come as you are" join the lab from the different computer science groups of the faculty. The primary focus is on combining natural speech and gesture recognition with high-end computer graphics for virtual construction tasks (Sowa, Fröhlich, and Latoschik, 2000). For simplicity, a toys set is used for demonstration. The architecture is an artificial intelligence approach where independent agents work together, based on predefined knowledge about the way the toys are constructed. These agents can be distributed across a network of computers. The VR Lab combines a Silicon Graphics Onyx^2 machine and several smaller computers. A stereo projection on a flat screen is driven by the Onyx^2 machine allowing one active observer to move freely in front of the screen and change perspective while viewing the scene with shutter glasses. Tracking units are attached to the shutter glasses and to the hand and arm of the experimenter. Additionally, the user wears data gloves which allow the tracking of changes in hand posture in situations where cameras would not be usable, for example, due to disrupted direct view. Gestures are used to disambiguate natural speech: for example, when referring to THE red bar and multiple bars are red. Humans naturally point towards the bar they refer to while speaking (Latoschik and Wachsmuth, 1998). The same system can be used for smaller construction tasks on a virtual workbench. All of the software is based on C and C++ code using Performer, AVANGO or directly using OpenGL.

NIH Resource Laboratory – Rochester

The VR Lab of the computer science department of the University of Rochester is funded by the National Institute of Health (NIH) and thereby open to many guest researchers. The projects therefore often have a connection to medical applications. In addition, researchers from the Center for Visual Science (CVS) in Rochester work in the same lab strengthening the connection to vision understanding. Psychophysical work in the context of visual memory and eye movements is combined with VR technology (Pelz, Hayhoe,

Ballard, Shrivastava, Bayliss, and von der Heyde, 1999). An Onyx Infinite Reality System with four processors and one pipe is the center of all VR applications in the lab. The visual scenery is presented via an HMD which is customized by monocular eye-tracking inside the helmet. In addition, position trackers are used to track head movements and to take pointing movements into account allowing interaction with the scenery. The projects range from EEG measurements during virtual driving on a hydraulic motion platform (Bayliss and Ballard, 2000), to computer models for driving. Performance of an autonomous virtual driving system and human observers was compared in a car following task. The system was based on the real-time processing of a video stream either coming from the SGI computer or natural sceneries captured by a video camera (Salgian, 1998). In virtual grasping studies Atkins, Fiser, and Jacobs (2000) examined the integration of vision and haptics using two large PHANToM devices. The equipment is programmed with C or C++ using Performer as visualization library. All equipment in the lab is connected to the main computer and the applications are thereby centralized.

Space Perception Lab – Santa Barbara

The Space Perception Lab at the psychology department of the University of Santa Barbara also uses VR technology. The main focus of the lab is around the perceptual and navigation behavior or abilities of humans. Recently, the integration of proprioceptive, vestibular and visual cues were studied (Chance, Gaunet, Beall, and Loomis, 1998). Several studies using a triangle completion paradigm (i.e., subjects were guided along two sides of a triangle and had to return directly to the origin) analyzed the impact of vestibular perception and proprioception on vision. In psychophysical experiments they compared real, virtual and imagined translation to examine the phenomenon of spatial updating (Klatzky, Loomis, Beall, Chance, and Golledge, 1998). Loomis et al. (1999) summarized the advantages and disadvantages of immersive virtual environments (IVE) in the areas of perception in general, spatial cognition for navigation, and social interaction. Their setup consists of two PCs: one for visualization and the other for data acquisition and optic and inertial tracking of human movements. Using an HMD allows the subject to freely move around in a certain area of the lab. The software combines C modules inside a php scripting environment. The VRML models are displayed directly by OpenGL functions.

Bankslab – Berkeley

The VR Lab at the University of Berkeley is mostly using visual cues in psychophysical experiments. They have recently begun to combine visual with vestibular cues. Their most recent addition is two PHANToM devices, to continue the research on visual-haptic cue integration in depth perception (Ernst, Banks, and Bülthoff, 2000). So far, the psychophysical studies used, for example, optic flow (simulated by 3D graphics) to determine perceived heading. Subjects judged whether the continuation of the path will be left or right of a target (Sibigtroth and Banks, 2000). The combination of optic flow with vestibular stimulation (Crowell, Banks, Shenoy, and Andersen, 1998) allows the study of self-motion perception during head turns. The use of a two axes tilt chair in combination with projecting the visual stimuli onto a flat screen (50°of horizontal field of view) is used to investigate flight illusions. The experimental technology is based on a PC system which allows, so far, no direct interaction and uses predefined paths. The visual rendering is scripted in matlab and directly uses OpenGL routines.

VR Lab – Umeå

The VR Lab in Umeå, Sweden, uses mostly visual simulations. In the past, haptic devices were used for interaction and led to the foundation of the ReachIn company. The VR projects range from medical applications to architecture and environmental planing. For example, a stroke simulator mimics the changes in perception suffered by stroke patients. Visual features can be changed and misaligned for the purpose of spatial neglect simulation. Another project in the medical field models the inside of the human mouth, including teeth. This application enables intense training for dental technicians for the design of prosthesis. A more traditional project simulates the inner region of the city. Other projects demonstrate the use of VR technology in general data visualization: Satellite pictures were combined with depth information to automatically derive a 3D model of certain parts of Sweden. A powerful Onyx^2 machine with ten processors and two rendering pipes enables stereo projection (with shutter glasses) in high resolution or the use of stereo HMD visualization. The models are built using Multigen and displayed via the Performer library.

1.2 Distributed Systems

Traditionally, computers are used without any connection to other computers. In the early 1980's the hardware for processing units, storage space, and memory became significantly cheaper. A number of groups had developed technology for local area networks using twisted pairs, coaxial cables or radio transmission. This network technology quickly became commercially available and enabled communities to distribute work across multiple, smaller processing units instead of using a big mainframe installation at a distant location. In the last couple of years, it has become extremely popular to connect computers (at least temporarily) to exchange data via the Internet.

Single units like computers easily run in parallel. They work on individual data sets and do independent work. In contrast, it is much more complicated to let two processing units work at the same time inside the same computer. The processing units of a multi-processor machine need to be synchronized in their work. Common data have to be updated between both units in order to guarantee a deterministic result for calculations. On the other hand most of the data is available to both units inside the main memory. The workload is equally distributed between multiple units by using scheduling strategies.

Having multiple units connected via a network without a common shared memory between them allows joint work on one problem only when messages are exchanged between the units in order to coordinate the distribution of work. The global status of the system is no longer known to each unit, but distributed across all members of the system. The distributed system is established through the communication between the units.

The general concept of **transparency**, which is defined as the concealment of implementational details and concepts from the user and the application programmer, is summarized, for example, in Coulouris, Dollimore, and Kindberg (1994). The most important forms of transparency are "access transparency" (access to local and remote resources with identical operations) and "location transparency" (access information of objects without knowledge of their location). Together they are sometimes called "network transparency".

1.2.1 Definitions and terminology

At an advanced workshop in 1980 LeLann summarized the objectives of a distributed system as follows (see Lampson, Paul, and Siegert (1981, chapter 1, G. LeLann: Motivation, objectives and characterization of distributed systems)). The **increased performance** of a distributed system in comparison to a multi-processor system is due to the inherited bottlenecks of the latter: shared memory, context switches, and interprocessor communication. The partitioning into several independent processing units allows parallel execution and asynchronous processing. The **extensibility** of the distributed system makes changes for performance or functional requirements possible. The simpler system design of the small units allows for easy installation and maintenance. Further, LeLann defines the term **n-resilient** as the amount of errors the system can experience without being disrupted in its functionality. The term therefore measures the **availability** of the overall system which is closely coupled to the **redundancy** of the single elements. Taking the term "resources" in a very broad sense, all these elements and also the data in the system can be shared in a distributed system. **Resource sharing** includes load sharing and the transparency implemented in the architecture. All the terms mentioned above are interconnected and can be applied to large networks, local area networks and multi-processor systems. In addition, the author states that none of the above objectives can be met without a central, **system-wide control** technique.

Single atomic operations of physically distant units of a distributed system will always have some non-zero time delay until the units can update and propagate their status. The global state of the system is therefore unknown to one single unit of the system. This introduces a real difficulty: the **synchronization** of the internal state between processing units. This is in stark contrast to non-distributed systems where the common status is known due to the **shared memory** concept (see Lampson et al. (1981, chapter 2, R. W. Watson: Distributed systems architecture model)). LeLann (Lampson et al., 1981, chapter 12, G. LeLann: Synchronization, page 282) offers a list of evaluation criteria for different aspects of synchronization. The criteria range from response time and throughput, extensibility, and determinacy to connectivity and simplicity. The relative importance of these issues for a given system depends highly on its purpose.

Research on distributed systems has progressed over the years. The collection of lectures in Paul and Siegert (1984) present tools and methods for **specifications**. The authors of the individual lectures follow the approach of the phase model for software engineering, starting at the level of acquisition and analysis, going through the specification of requirements and continuing with the design of the system architecture. Further, the design of components and their specification crystallizes into the integration and installation of the system. In addition, the authors introduce language constructs and paradigms for distributed programs.

The use of **object orientation** in the context of a distributed operating system is discussed, for example, by Horn (1989). He uses the example of a system designed to handle electronic mail to examine different strategies. The client-server concept is contrasted with distributed shared memory especially for the question of where the object is located. The client-server mechanisms might be too heavy-weighted to transmit a simple text messages but certainly would fit the requirements. The overhead of object structures might reduce performance. Shared memory, on the other hand, is considered to be the alternative and complementary strategy. Rather than sending the request for working on the data in a remote procedure call, the data itself is transfered to the node which is currently working on the data. Further, the persistence of objects in a distributed environment is discussed. The possibility of taking snap-shots of the current state of the whole system and continuing at that point later is very often coupled with features which allow the activation or deactivation of single objects. Tools that allow general object definitions which can be handled with different programming environments and languages remain a challenge.

All of the above can in principle be implemented on different abstraction levels. In an UNIX environment the file system already provides distributed access to files and other resources. In a database system, client-server concepts implement some of the above features by means of providing data in different locations, and care for the coherence and deterministic behavior. Others have implemented distributed applications making use mainly of the performance advantage of multiple processing units. To be more specific, the next section will introduce a few examples which realize these distributed concepts.

1.2.2 Examples of systems and their application

Examples of distributed systems can be found at the level of operating systems (OS) and protocols all the way up to specific applications. In all distributed systems one or more of the following features are distributed across several nodes or computers: **resources, workload or data**. Today's biggest distributed system is the World Wide Web (WWW). Most of the computers today are at least temporarily connected to the Internet. This network is used to distribute data usually representing text, pictures, video and sounds. The WWW is a combination of several protocols which all work on top of the basic Internet Protocol (IP). These additional protocols (http, ftp, telnet) enable different services and have their origin in the UNIX environment. Smaller networks of computers, so called local area networks (LAN), are often more closely coupled in the sense of a distributed system. They share resources like printers and disc space. Operating systems often enable the sharing of data inside the network as a network file system (nfs). The single data files are available for direct access by all the computers inside the LAN. The OS therefore has to carefully control the file access. For example, writing permission is always granted to one process at a time inside a network to guarantee consistency of the data.

The distribution of workload can again be implemented on different levels. Operating systems, programming languages or protocols can be used to let multiple units work cooperatively together. Three methods of distributing the work can be identified: message parsing, shared memory or remote procedure call (RPC). Very often message parsing is the basic concept with which the other two are implemented.

Based on the application and goal of the specific project, there are many distributed systems approaches which can be found in the literature. The summary by Cheng (1993) characterizes roughly 100 different systems implemented on different levels as indicated above. Only few are discussed here to give brief examples for different approaches.

One system which is used to distribute workload across a network is the Parallel Virtual Machine (PVM) (Geist, Beguelin, Dongarra, Jiang, Manchek, and Sunderam, 1994). The system was developed to use a heterogeneous network of computers which was connected to one virtual machine with one programming interface. More than 40 different OS are supported by the library which is free for all UNIX derivates. The communication between the nodes of the network is realized via message parsing inside the PVM daemon which needs to be running on every system participating in the virtual machine. This system is very often used to implement algorithms with a high degree of parallelism. Other systems for parallel computation use the PVM as a base system and implement a separate application layer.

A more recent approach to distribute data and functionality across a network, to host different applications, and to control the flow of information at the same time is the Distributed Application's Communication System (DACS). It is described in detail and compared to other distributed systems in Jungclaus (1998). It was developed in the context of the SFB360[1], a project of several work groups focusing on linguistic and cognitive mechanisms in human-machine communication in the scenario of a construction task. This task was already mentioned in the context of the VR Lab in Bielefeld (see section 1.1.3). The individual parts of the overall projects had to be flexibly connected. The DACS offers multiple communication primitives (streams, RPC's and messages) with a simple interface. Similar to the PVM, on every system used by DACS a daemon must run to deliver messages and start new programs on request. Further, DACS is limited to UNIX like OS architectures where preemptive multitasking is available[2].

Other systems are implementing distributed games across the Internet (Frécon and Stenius, 1998; Harada, Kawaguchi, Iwakawa, Matsui, and Ohno, 1998; Powers, Hinds, and Morphett, 1998; Greenhalgh, 1998; Wray and Hawkes, 1998; Reitmayr, Carroll, Reitemeyer, and Wagner, 1999). The particular problem is to update the status of the game via slow Internet connections which cause high latencies. In general, these distributed systems solve this problem by partitioning the environment either statically or dynamically based on the near surrounding of the the individual players. Some of these systems pretend to deliver Distributed Virtual Environments (DVE). Coming back to the definition of Virtual Reality and Virtual Environments, these systems do not qualify since only descriptions of visual scenes are shared and rendered asynchronously on different machines. The systems are limited in performance and do not include touch, auditory cues or other sensory modalities besides vision. The level of interaction is often limited to text input or movement control of the player's simulated body or avatar.

Another system which aims at desktop VR with multiple viewers of the same virtual scene is described by Demuynck, Broeckhove, and Arickx (1998). They classified virtual objects and participants as so called "entities" in their network. Interaction between entities is done by message transfer. The authors evaluate different peer-to-peer protocols with respect to their throughput and latency. To ease bandwidth demands, the system divides the environment into smaller parts (grouping). The paper describes only a prototype of the system which does not yet include load balancing or a potential port to WindowsNT.

Other approaches call themselves distributed second generation VR systems, as part of an attempt to distinguish themselves from the first generation approaches, which were mostly ad hoc implementations. Bangay, Gain, Watkins, and Watkins (1997) defined the characteristics necessary to be second generation as: configurable, parallel, distributed and providing Virtual Reality hardware support. They presented a modularized system that uses several I/O devices (Polhemus trackers and data gloves). Based on a message parsing of a point-to-point communication, a distributed object database and a virtual shared memory architecture is implemented. So far, the system simulates visual stimuli only. Therefore, in the light of our definition (see 1.1) both systems (Bangay et al., 1997; Demuynck et al., 1998) do not qualify as VR due to their usage of too few modalities.

[1]Sonderforschungsbereich 360: Situierte Künstliche Kommunikatoren

[2]Specifically, DACS does not run on Windows.

Some authors have extended normal desktop multimedia applications to run on several PCs in parallel (e.g., Husemann, 1996). This approach is concerned about delivering synchronous video and sound replay on distant machines. Husemann presented a case study of video on demand as a proof of concept. Others use the degree of interactivity to make a clear distinction between Distributed Virtual Reality (DVR) and Distributed Multimedia (DM). Wedlake, Li, and Guibaly (1999) call DVR a form of DM. To analyze current systems, they propose five stages between operating system and application level. Their approach explicitly splits application layer and programming system layer to gain flexibility and a general application programming interface (API) for DVR.

1.3 Human psychophysics and physiology

This section is meant to introduce the reader to general concepts and important definitions as far as they concern this thesis. After some basic terminology, some results from the literature serve as examples and for future reference in the experimental chapters. This very brief introduction can not replace intense and deep knowledge about human psychophysics and physiology.

1.3.1 Definitions and terminology

Classical psychophysics

Psychophysics defined as the description, quantification, and interpretation of perception, goes back to G. T. Fechner who established the rules for psychophysical experiments (Fechner, 1860) with three demands:

- The stimulus condition has to be **controlled** for all aspects.
- Each individual condition has to be **repeatable**.
- To measure the influence of a physical stimulus dimension, the stimulus must not be varied in more than **one dimension** at a time.

The basic idea of psychophysics is therefore to measure the dependence of a subjective reaction on certain physical properties of a given stimulus sequence. Varying the stimulus property in a physical dimension can change the subjective percept and result in different behavior from the organism.

The two primary psychophysical values measured are **absolute threshold** and **just noticeable difference (JND)**. The response of a sensor, which is normally related to a certain percept, can be differentiated into three main parts:
A) The stimulus is too weak, that means below threshold and the sensor does not react. B) The stimulus is too strong and the sensor reacts with its maximum output due to saturation. C) In the interval between A) and B), the stimulus' intensity is appropriate for the sensor. In this range the sensor very often shows exponential or linear response properties. Further, the JND ($= \Delta i$) increases with stimulus intensity i. The ratio between them is constant over a certain range which is known as **Weber's law**: $k = \frac{\Delta i}{i}$. A related functional dependency can be described by **Steven's power law** as: $Y = kX^n$ where X is the physical property, Y the measured perception, and n is calculated as the correlation coefficient of $logY$ being the estimate by $logX$ (Wiest and Bell, 1985).

Human physiology

Humans are traditionally considered to have just five senses: sight, hearing, touch, smell, taste. A seldom mentioned sixth one is the vestibular sense of body accelerations. The human body often combines several organs to sense different physical properties of given stimuli. The traditional touch for example is in fact a combination of temperature, pressure, deformation, and vibration sensors. To be more specific, we define different senses in combination with the physical property they are sensitive to. The adequate stimulus defines thereby the sensor.

The human eyes (more specifically the cells of the retina) are sensitive to photons with wavelengths between 400 and 700 nm. The perception of light is called **vision**. Visual stimuli can contain information which is perceptually different. On one hand, the main factors are color (wavelength) and luminance (light intensity). On the other hand, the spatial and temporal aspects of the pattern of light hitting the retina is important. The spatial pattern contains contrast between regions, preventing the visual stimulus from appearing homogeneous. Contrast changes can, for example, define edges or regions which belong together. The pattern of light is perceived over time with similar properties. Changes over time are most commonly connected to moving objects in space[3]. One example is the sequence of rapid changes in the brightness of multiple similar stationary objects perceived as apparent motion of one object.

The **vestibular system** (also known as labyrinth) allows the sensation of self-motion in six degrees of freedom in the absence of other external cues (like vision). It consists of two subsystems combining linear and angular acceleration sensors: the canal system for angular acceleration and the otolith system for linear acceleration (Wilson and Melvill Jones, 1979).

The **canal system** consists of three nearly orthogonal circular tubes filled with viscous liquid (endolymph). The cupula is deflected inside the liquid due to angular head accelerations. Because of mechanical properties of the cupula-endolymph system, the sensors encode angular velocity. This could be mathematically seen as the integration of the acceleration over time which becomes imperfect for frequencies below 0.1 Hz (Mergner, Nasios, and Anastasopoulos, 1998).

The **otoliths** are small crystals embedded in gelatinous mass around sensor cells. They are influenced by linear acceleration, due to the density difference to the surrounding medium. The otolith system therefore codes gravitational direction as well as linear acceleration from head translations. The compound signal is the superposition of both. This otolith afferent signal is the input to the CNS (central nervous system) which solves the gravitioinertial-force problem using internal models (see Merfeld, Zupan, and Peterka, 1999).

The human body is mechanically controlled by muscles. Each muscle has sensor cells called muscle spindles which monitor the contraction of a muscle. Other cells inside the joints sense pressure and thereby code flexion of the body joints. The combination of the perception of muscle spindles and joint flexion receptors is called **proprioception**. The current mechanical status of the body is therefore known from proprioception and by the motor command which last caused any changes (**efference copy**).

[3]Otherwise brightness changes are perceived.

The skin of the human body contains several sensor populations with changing density across the whole body. Some of them are sensitive to pressure (e.g., Merkel cells) others to vibrations (e.g., Vater-Pacini cells). The combination of all skin sensors (pressure, temperature, pain, vibration) and proprioception is summarized as **somatosensoric** information.

1.3.2 Psychophysical methods

In order to psychophysically examine the human sensory system, different methods can be used. Common for most methods is that a judgement is given by the subject as an answer to a new stimulus condition. Experimental paradigms make use of reaction time as well as the judgements themselves. So called two alternative force choice (2AFC) experiments present stimuli and ask a specific question forcing the subject to answer with one of the given alternatives. Signal detection theory is often used to interpret the results. Hits (the correct answers identified by the subject) and false alarms (positive, but wrong answers) are compared with misses (correct answers not identified by the subject) and rejections (correctly rejected answers). The sensitivity measure d' indicates the ratio between the frequency of false alarms and misses to judge whether the information provided was correctly used by the subject. This method can use simple yes/no paradigms to identify perceptual thresholds or to draw psychometric functions providing information about the sensitivity of the system to a given physical stimulus dimension. Other judgement methods are discussed together with relevant applications later in this section.

Threshold detection

In the context of this thesis, some visual and vestibular thresholds are of special interest. In order to simulate appropriate stimuli for different modalities, the thresholds for those modalities should be known. However, before presenting stimuli it has to be discussed in which physical dimension the threshold should be searched for. For example, measuring luminance thresholds of a colored patch will highly depend on the color used. Rods are not able to distinguish colors, but have a higher light sensitivity than cones. Does it therefore makes sense to judge luminance thresholds for colored patches, or would it be better to use grey patches?

We face a similar problem when measuring thresholds for the vestibular system. The canal system is classically known to respond to accelerations. Nonetheless, it remains unclear if thresholds for velocity, accelerations or even jerks can be transformed into one threshold for the system. Mergner, Siebold, Schweigart, and Becker (1991) have shown that there is a "velocity threshold" of the order of 1°/s in experiments involving interactions of neck and vestibular stimulation. Others claim that "vestibular response thresholds, latencies and amplitudes appear to be determined strictly by stimulus jerk magnitudes. Stimulus attributes such as peak acceleration or rise time alone do not provide sufficient information to predict response parameter quantities." (Jones, Jones, and Colbert, 1998). Both results contradict to the degree that a velocity of 1°/s can be reached with multiple jerk profiles having a different maximum jerk. Will those profiles be perceived differently and will some cause the sensation of self-motion and others will not?

Other problems occur in the definition of thresholds when different modalities interact. For example, the detection threshold for earth relative object motion depends on the current self-motion of the observer (Kolev, Mergner, Kimmig, and Becker, 1996). The threshold for perceiving self-motion is therefore relevant for the visual perception or at least for the visual interpretation of the environment. Other researchers used estimates of threshold and developed a model of how optic flow is used by pilots during landing phase. Different aspects of visual optic flow were identified by comparing real flight trajectories with psychophysical threshold data (Beall and Loomis, 1997).

Judgement and estimation methods

To judge thresholds accurately, the **method of limits** is often used. Stimulus intensity is increased from a value where no perception can be observed until the subject reports reliably the existence of the stimulus (upper boundary). From there, the intensity is reduced until the subjects reports the stimulus feature has faded (lower boundary). The threshold is assumed to be at the mean of upper and lower limits. The differences between the measured limits is due to the so called hysteresis of perception. This technique is, for example, used for threshold detection for angular velocity for the vestibular system (Kolev et al., 1996).

Some methods are known to work best for judging the magnitude of a stimulus. However, one has to be careful about the kind of estimate one asks for. Poulton (1981b) made the important suggestion that one should distinguish between scales which are known (inches or meters) and new scales one is not familiar with (like [g] - as earth gravitational force). The judgements on an unknown scale will follow a logarithmic stimulus spacing whereas the judgement on a well known scale is linear, if numbers with roughly the same amount of digits are used. A judgement on the scale 100 to 1 is likely to result in a linear stimulus spacing.

Stevens' magnitude estimation (Stevens, 1957) is described with other methods by Poulton (1968). It was used, for example, to judge angular turns and maximum peak velocity in Mergner, Rumberger, and Becker (1996). Multiple standard stimuli are used for anchoring, and new stimuli are judged as multiples or fractions of the previously presented standard stimulus. This method is often used as magnitude estimation where the absolute scale is not important. Relative scaling normally results in a linear scaled inter-stimuli distances for known modalities.

Other estimation methods display a target and let subjects track the change in self-motion by anchoring a handheld pointer on the distant object. This method was used to measure the perceived changes in position and orientation in space by Ivanenko, Grasso, Israël, and Berthoz (1997b). In other studies, subjects directly walked towards the target (Glasauer, Amorim, Vitte, and Berthoz, 1994) or reported when they passed the memorized target (Harris, Jenkin, and Zikovitz, 1998, 1999).

Stevens exponent and range effects

S. Stevens claimed that the perceived stimulus which can be described with an exponential function is a fundamental feature of the specific sensor. The exponent of this function should be unique within a given species.

In a meta-analysis, Wiest and Bell (1985) compared the Stevens's Exponents measured in 70 **distance judgement** studies. They summarize the main points as: A) The exponents were on average close to 1.0 (mean = 0.95, median = 0.98). B) On average, the simple linear regression results in a slope of 0.85, indicating that a general 15% underestimation occurs. C) Comparing a simple linear regression with the results of fitting Stevens' power law reveals that on average only 2% more of the variability is explained by the power function.

Others claim that Steven's exponent is mainly chosen by the experimenter and determined by the ratio of maximum and minimum distance (Poulton, 1967). Kowal (1993) describes this as an inverse relationship between stimulus range and the exponent in Stevens's power law. Others have found range effects especially for judging angles and distances (Ivanenko, Grasso, Israël, and Berthoz, 1997a; Berthoz, Israël, Georgesfrancois, Grasso, and Tsuzuku, 1995).

1.4 Thesis overview

The end of this chapter introduces the main ideas of the thesis, guiding the reader through the remaining chapters. Up to this point, the general introduction of central definitions and a description of other VR labs were given. In the next section, the Motion-Lab at the MPI for Biological Cybernetics will be introduced. Afterwards, the central questions for the work in the Motion-Lab are described. To answer these questions, two experiments are proposed with their detailed goals. Finally, the summary will review the achievements and present the main results.

1.4.1 Motion-Lab

The Motion-Lab will be described in chapter 2. Is was constructed at the MPI in the Spring of 1999 and has been in development up to now. The hardware was installed during the Summer of 1999 and completed during the Spring of 2000. Since then, experiments have been conducted to address the questions described in the next section. Certain design decisions have influenced the whole lab. Chapter 2 points out the main design criteria and introduces the solutions which were implemented. Open and closed loop experimental conditions are discussed along with their influence on the software design. The Motion-Lab is then described, giving details about existing hardware and software. The Motion-Lab is an example of a distributed VR system, since the simulation of different modalities is distributed across a local network of standard PCs. While the lab has so far been used to conduct psychophysical studies on spatial updating, chapter 2 ends with a description of other possible applications.

1.4.2 General Experimental Questions

The primary question we have explored in the Motion-Lab concerns the perceived location of oneself in real space as well as in a virtual environment. This location (here including position and orientation) is important for the interpretation of the other senses. For example, perceiving our location enables us to disambiguate between possible interpretations of a visual scene.

Normally, we know our location and it is naturally updated when we move through space. How is this updating related to our perception? Which of our senses contribute to this automatic spatial updating? If some senses, for example vision, do not contribute, but would profit from the update, a strong coupling of several modalities in our perception would be the result. In traditional psychophysics, one specialized cue in one modality is studied (for example, color perception in vision). However, more recently psychophysics was extended to look for cue integration and adaptation effects across modalities. Nonetheless, no general model so far explains the network of influences between our senses. Two experiments in the Motion-Lab will be presented to explore inter-modality effects which allow us to speculate about general features of sensor fusion.

The first experiment (chapter 3) asks for the general dimension of the perception when moved in the dark. Specifically, is the distance, velocity or acceleration directly perceived or does one derive estimates of those values? The sensor organs (otoliths and canals of the vestibular system) are certainly stimulated by angular and linear accelerations, respectively. But are those accelerations transformed, mathematically integrated, into a velocity estimate? Further, is the velocity value, if it exists, usable for integrating a second time to come up with an estimate for traveled distance or angle?

In order to perceive our environment as stable during movements, we have to stabilize ourselves, too. The second experiment (chapter 4) focuses on the question how well we can stabilize ourself in space and learn certain characteristics of a path. Specifically, can we code the angular amplitude (heading turns) in space during a task where we follow a virtual path without actually seeing it? Does the path following allow us to learn the path and repeat the turns we learned? In the experiment, we focus on two cues that provide no absolute spatial reference: optic flow and vestibular cues. Specifically, we asked whether both visual and vestibular information are stored and can be reproduced later. The experiment therefore tries to dissociate which information (visual or vestibular) is used for the memorized path. Further, are those modalities integrated into one coherent percept or is memory modality specific?

Both experiments are connected by the question of how we perceive turns. In the first experiment, verbal judgements about the heading changes will be compared with linear movements. In the second experiment, the turns are presented in multiple modalities and are tested in a cue conflict condition. Finally, the integration and interaction of multiple senses is discussed.

1.4.3 Main results

The main results and achievements will be summarizes in chapter 5.

The presented Virtual Reality setup uses a distributed network, but hides this effectively by providing a client/server architecture. Several device servers and the corresponding clients are implement in the Motion-Lab programming library. The communication is asynchronously done in combination with a predictive algorithm reducing the latency in the system. C++ classes make the lab easily accessible for the programmer of an experiment.

The functionality of the Motion-Lab will be demonstrated with two experiments. The results from the first experiment shows that humans can judge their spatial location based on vestibularly perceived distance and velocity. The judgements of maximum acceleration were similar to the ones for maximum velocity, indicating that acceleration is not easy to judge for most people. The second experiment will allow us to interpret some interesting interaction in the fusion of the visual and vestibular perception as a “max-rule”: The perception of the modality which appears to change most is used to reproduce turns from memory. This result will be discussed in the context of modality specific or distributed memory.

Chapter 2

Motion-Lab

This chapter documents the general ideas and implementational details of the Motion-Lab. This lab combines VR equipment for multiple modalities and is capable of delivering high-performance, interactive simulations. Crucial design decisions are explained and discussed as the software and hardware is described. The goal is to enable the reader to understand and compare this implementation of a distributed VR system with solutions demonstrated by other labs as described in the introduction (see section 1.1.3).

The overview starts with a general discussion of VR systems as simulations for multiple modalities. This is followed by a short discussion of the advantages of distributed solutions in contrast to a mainframe realization. Focusing on the distributed system, basic communication problems are mentioned and one approach for the solution of those problems is introduced as the main communication structure in the Motion-Lab.

The hardware section (see 2.2, p. 26) demonstrates the variety of hardware used in the lab and introduces all the devices to the reader. Each piece of equipment is described in terms of its functionality as well as the technical details. Alternative solutions are discussed and the main differences are rated.

Finally, the software concept is described in detail, but without going too deeply into the source code (see section 2.3, p. 38). The latter is available online[1] in combination with the DOC++ documentation of most of the parts. General ideas about software development which guided this project are compiled into a short introduction to distributed software development in context of multiple OS.

2.1 Overview and purpose

This introduction to the realization of a distributed VR system explains some of the general design criteria and concepts. Different principles are discussed and guide the reader towards an understanding of the overall system. This part is meant as an introduction to the Motion-Lab for the purpose of this thesis, as well as a guide for those who start working in the lab and would like to learn basic rules and principles.

[1] http://www.kyb.tuebingen.mpg.de/bu/people/mvdh/motionlab/source

2.1.1 VR systems integrate simulations for multiple modalities

As the reader has seen in the introduction, the realizations of many so called VR systems, are confined to the simulation of a visual world. Most of the setups involve at least one interaction device for controlling a virtual camera, allowing the observer to change the view. Nonetheless, our VR definition given in the introduction (see page 1) requires the involvement of more than one modality in the simulation. Some authors like to call the input device itself a device for haptic interaction just because one touches it. Very rarely is force feedback provided in real time for the controlling devices and therefore the information is often going only in one direction: from the user into the system.

In driving simulators, acoustic cues are relatively simple to add and control. Starting a virtual car and changing pitch of the motor noise or simulating other sound properties with respect to the driving parameters like speed is adding to the sensation of a realistic system. Background auditory stimulation has been shown to considerably improve the sense of presence (Gilkey and Weisenberger, 1995). Providing sound which is simulated in three dimensions is more complicated and involves considerably more effort. However, the sense of presence in the VE is significantly enhanced by spatialized sound in comparison to non-spatialized sound as Hendrix and Barfield (1996) pointed out.

Flight and driving simulators can be divided in two groups by considering vestibular cues. Some of the simulators are mounted on so called *motion platforms* to be moved as whole. The others can not simulate whole body movements (by means of short accelerations) and are called *fix-base simulators*. In both cases, simulations try to move an observer in a large virtual environment. Nonetheless, the simulators themselves stay in a confined space even when they can move a short distance. The mismatch between large changes in simulated location (movements in three dimensions) and the actual position in the laboratory might be one factor of simulator sickness (Viirre, 1996). The real accelerations can not possibly be matched to the simulated accelerations without performing the actual perfect movement. The movement type which should closest approximate the important information for the vestibular system of humans is defined by terms of *motion cueing* or *motion simulation*.

In the Motion-Lab at the MPI, visual, haptic, vestibular and acoustic simulations are integrated into the system. Subjects[2] can therefore perceive a simulated world in many different modalities. The simulation of non-spatial sound is clearly the most simple one, due to limited implementation time. On the other hand, it provides ways of instructing the observer even without vision by a synthetic speech system presented via headphones or loudspeakers. The other modalities involve additional hardware equipment which is often accompanied by software and libraries from the manufacturer. Integrating different modalities requires the design of control programs for the specific modality based on different time limits. For example, a visual display runs optimally at a rate of 60 Hz, whereas haptic systems should reach beyond 1 kHz and sound simulations should be even faster (44 kHz to reach CD quality). Showing this wide range of speed requirements, it makes sense to work the devices in a parallel manner, not disturbing each other.

[2]Persons participating in the experiment are in the following referred to as masculine or feminine. It is understood, that the respective other gender is meant to be referred to as well.

2.1.2 Distributed system or stand alone computer?

There are mainly two distinct ways of achieving powerful simulations:

- The "big solution" runs on one very fast mainframe computer providing all the connections to different parts of the equipment.
- The distributed solution connects smaller and specialized computers which are connected to one device at a time, but provide sufficient speed to run this device at the required speed.

These two solutions are discussed by focusing on their respective qualities and drawbacks in the two following sections. The advantages of one solution are very often the disadvantage of the other.

The "big solution":

One of the obvious but important advantages of having one big computer which does everything in the VR-setup is that no synchronization between multiple databases for different modalities is necessary[3]. In this solution there is just one OS and one set of libraries involved. The maintenance is therefore low in sense of work for a technician, but the costs for the special hardware which might be involved are considerably high. Furthermore, an advantage which clearly separates this solution from the other is the possible load-balancing across modalities. The programmer can design his program in a way that the most urgent work is done first. In addition, the "big solution" can reliably synchronize tasks below a time resolution of 1 ms.

The biggest disadvantage might be the missing support for some special hardware with a given OS. It seems to be difficult or at least much more expensive to get some parts and interfaces changed later. The system mostly stays as it is because it is hard to extended only a part of it. The overall costs are considerably high, since the computer needs to be in the high performance sector of the market. Special cooling and noise problems might occur and additional problems are posed by short cabling or other interface communication.

The distributed solution:

There are some advantages of the distributed solution which could at the same time be seen as disadvantages of the "big solution". It is more flexible and easy to extend the system gradually or to substitute parts of the whole system. The use of special OS platforms and special libraries for those platforms becomes possible, since not all the different computers have to run the same OS. For most of the parts, it becomes possible to use standard components, which are more common and have the advantage of lower investment costs. Different parts of the simulation – or let's say the system – are running independent of each other which makes critical parts safer from general crashes of the system. The stability of the overall system increases since the nodes of a computer network can replace each other in functionality.

[3]Nonetheless, most VR programs still use different object trees for different modalities. However, there are recently several approaches which try to integrate, for example, sound, haptic and vision into one representation.

The biggest disadvantage is the communication overhead of the system for synchronizing the data. No real synchrony is possible, but if the speed of the system if sufficiently high, the reached synchrony is acceptable for some projects. Ryan and Sharkey (1998) propose a smooth connection between asynchronous update for objects close to the observer, allowing real-time interactivity, and synchronous, but delayed update for distant objects. The authors argue that network latency (differences in time) thus will not cause discontinuity in space for the user.

2.1.3 Distributed components and asynchronous communication

We decided to implement the Motion-Lab as a distributed VR system. The following section will explain how the communication latency is actually made acceptable for our system by implementing a "soft synchrony" strategy. The communication is the crucial point of a distributed system, especially when different processes have to provide fast feedback at different speeds. The main point of the communication in the Motion-Lab is the asynchrony of all processes and the explicit statement that there is no guarantee for a special message to be accepted by the recipient at a given point in time. Moreover, the information has to be coded in a way that provides the current state and additional information which allows the recipient to extrapolate the status into future.

Let us illustrate the main problem with an example. Imagine the situation where an input device (e.g., a joystick) is controlled by interrupts on a system level and therefore has a rate between 1 and 60 Hz. On the other side, a motion platform is updated very strictly with 30 Hz. The simulation in between has to connect to both devices and the programmer chooses to run it at 10 Hz. In some of the simulation steps, there is no new input from the input device, but the system just takes the last known value. In other simulation steps several records of the input devices had been available, but the last record is the most important, since it codes the most recent state. For the simulation it might be sufficient to always take the last known record to update, for example, an internal model to move a virtual observer forward. Based on the internal state, the simulation can therefore send information to the motion platform, which would arrive there at a rate of 10 Hz. The platform needs to interpolate now for at least two steps between two new data records in order to come up with a smooth movement at 30 Hz rate.

There are two opposing principles working here: One to slow down update rate (from 60 to 10 Hz) and the other to interpolate in time to increase update rate (from 10 to 30 Hz). If the process which is providing information is running at a higher speed (faster update rate) than the consuming process, it is always safe to take the last record which was available. Having the situation the other way around would then result in a jumpy movement and would cause noticeable disturbance for visual and vestibular simulations. Therefore, if at some point the consuming process is running faster than the data records from the providing processes arrive, the program should extrapolate from the last know record into the future to guess the momentary status of the system. If at each point in time the status of the system (take position of an observer as an example) is known, together with a prediction of the rate of change (velocity), extrapolation becomes easy. This extrapolation method is also useful when short breaks in the system make the information flow unsteady and change the update frequency. Since a lot of different devices work at their inherent speed or changing rates, it makes sense to soft synchronize them using the above principles (see Fig. 2.1).

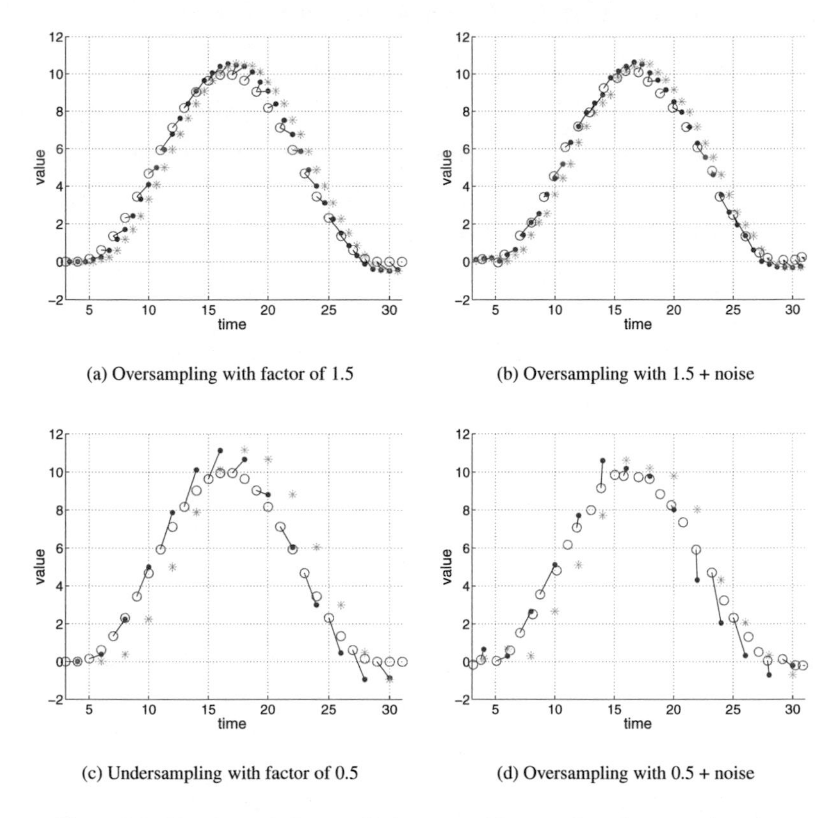

(a) Oversampling with factor of 1.5

(b) Oversampling with 1.5 + noise

(c) Undersampling with factor of 0.5

(d) Oversampling with 0.5 + noise

Figure 2.1: The extrapolation method can provide smooth paths even when the provided data flow is not smooth and in addition unsteady in time. Two examples show that the extrapolation works for under and oversampling relative to a given frequency. The left panels show data results for two frequencies without noise. The right panels show for the same frequencies the results for a situation where time and data are overlayed by random noise adding +/-25% of the respective units. The one dimensional case can be generalized to serve all six degrees of freedom for camera or motion-platform data. The red circles indicate the transmitted position data on which bases the last velocity was calculated. The blue dots indicate the derived position connected with a thin black line to the data point they are based on. The green stars indicate a potential rendered image based on the blue position at the time the image would be displayed. Unsteady and changing frequencies normally cause “jumps” in continuous data when the stream is re-sampled with a fixed rate. In contrast, this extrapolation method predicts a future point based on position and velocity information. Even sudden changes of rate (due to incomplete data, for example) will not disturb the smoothness of the data. Due to the velocity prediction the algorithm overshoots, displaying behavior similar to low-pass filters.

2.1.4 Synchrony, and closed- or open-loop functionality

Synchrony is an issue by itself, when different modalities are involved. If someone drops a cup and one hears it breaking on the floor before seeing it happen, the situation would seem unnatural to us. Having the sound reach the ears later than the visual event reaches the eyes would, on the other hand, feel normal when seen and heard from a distance, since sounds travels more slowly than light. Extending the distance further, one would always expect to perceive the lightning before the thunder. Events in the real world often provide feedback in multiple modalities. If we substitute some of the modalities in the simulation with a virtual version, we should provide the same synchrony. Exceptions are experimental paradigms explicitly dealing with time differences as done by Cunningham, von der Heyde, and Bülthoff (2000b)[4]. In a closed loop experiment it will therefore be necessary to provide synchronized feedback for events which were caused by the observer without a noticeable loss of time. The feedback should be provided in a closed loop so that every action is directly coupled to its effect (see Fig. 2.2).

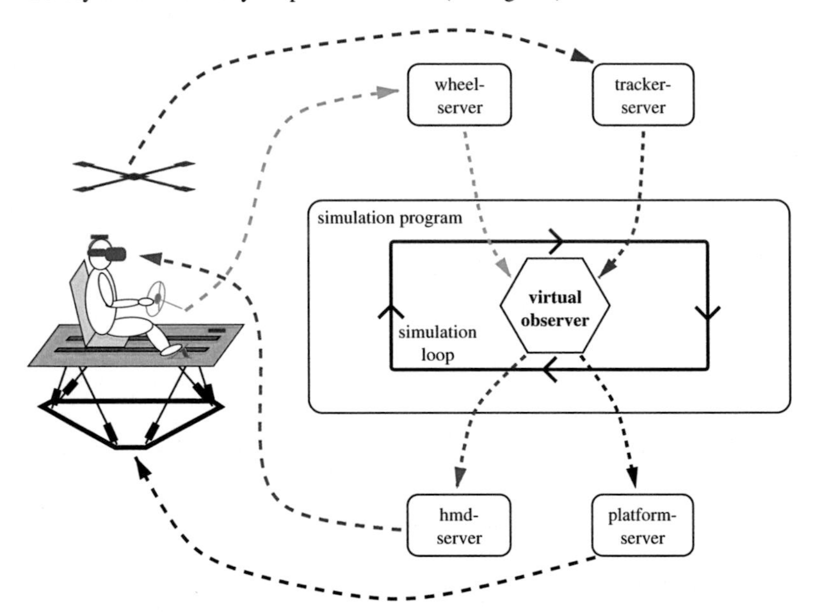

Figure 2.2: The closed loop simulation feeds back the actions of an observer to the modalities he experiences in the simulation. Every action is coupled to the reaction in the simulated world.

For an open loop condition the observer has no influence on the occurrence of events in time[5]. The system gets simpler when it does not provide feedback to the actions of the observer (see Fig. 2.4): The simulation can be reduced to a playback for different modalities in a predefined time schedule. If the accuracy of time resolution and synchronization on

[4]Even then we have to know the exact point in time of certain events in order to add additional time offset in the program.

[5]Note: If the simulation provides no feedback to actions, it does not fulfill the given requirements for VR!

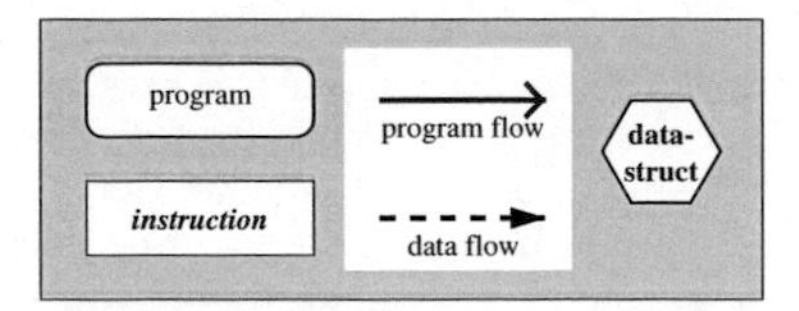

Figure 2.3: Legend for all flow figures (2.2, 2.4, 2.8, 2.11, and 2.19).

a system level is guaranteed, the playback will appear synchronous to the observer[6]. One could, in this situation, exactly define events to occur at a certain point in time and compensate even for slow transfer rates, if the simulation is completely known beforehand[7].

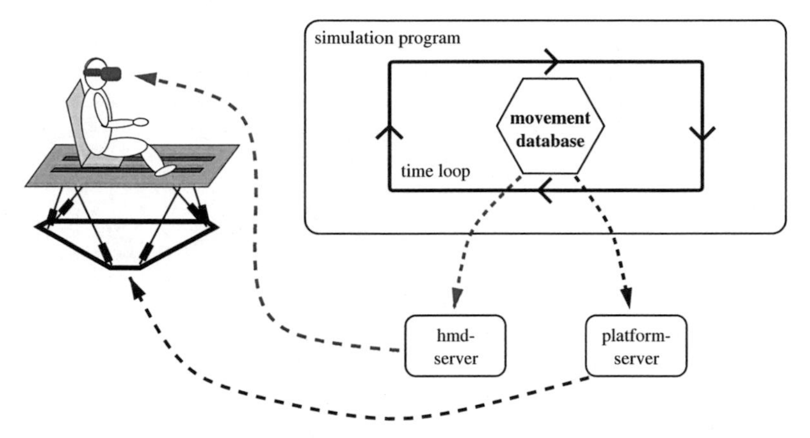

Figure 2.4: The open loop condition is in comparison easier because most of the events in the simulation can be calculated beforehand, stored in a movement data base, and do not have to be updated based on the users action.

In contrast, the closed loop condition demands that the system reacts with respect to actions of a observer. The level of interaction determines the level of feedback required to let the simulation appear realistic. For example, in a simulation of a race car the steering wheel is the primary interaction device. If the wheel is providing the steering angle to the simulation, the camera could be updated simulating a moving observer. Driving in this simple simulation of a car does not feel real; the "sense of being there" is quite low. Adding force feedback centering to the steering wheel would improve the feeling of driving a real car. Furthermore, the driver can tell from the haptic feedback alone whether he is going straight. Extending this idea even further, the force model of the steering wheel could include the speed of the car: turning the tires on the spot would be harder at low speeds and so on. The information provided by the combination of both modalities (vision and haptics) is similar, but the coherence and synchrony makes the simulation more realistic. Including sudden jerks when leaving the road or velocity coupled noise, for example, would add information which could not be visually perceived. This example offers

[6]Ignoring processes which have order effects and take history into account.

[7]A lot of "fun-rides" in Disneyland, for example, are well predefined and worked out for play back. If decisions of the observers are taken into account, alternative outcomes are defined and the observer works down a tree of decisions.

no systematic proof that synchrony and coherent feedback in different modalities make a better driving simulator. However, it makes a plausible suggestion of what could be gained by having those features. So far, there has been some evidence that presence, the "sense of being there", improves task performance in VR (Witmer and Singer, 1998). However, as Witmer and Singer pointed out, presence is negatively correlated with simulator sickness, which leaves the direction of causality between the two unclear.

2.2 Hardware

In general, the hardware used is standard commercially available equipment, with the exception of the force feedback steering wheel[8]. For later ease of reference, the next sections include short descriptions of the different devices and their basic working principles and functions. The information is mostly provided by the manufacturer, but is all rephrased and simplified for the purpose of this thesis. Specific questions referring to technical data or functional details should be directed to the addresses given in appendix D. A general overview of the Motion-Lab equipment is given in Figure 2.5.

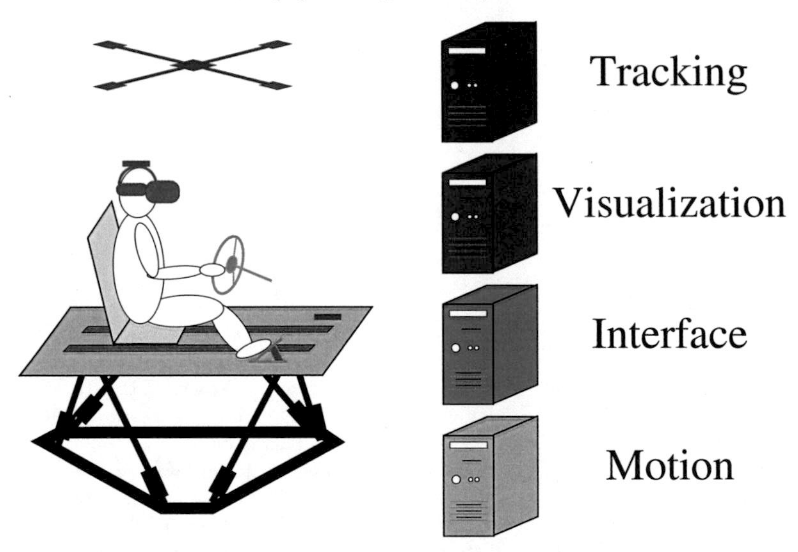

Figure 2.5: The Motion-Lab setup in its main parts consists of the motion platform with a seat for a human, interfaces for him/her to control his/her position in the virtual world, the visualization of that world presented by an HMD, and a tracking system to render the corresponding viewing angle for the measured head position. Each device is controlled by a separate computer to guarantee optimal performance.

[8]It was designed and constructed in the institute's own workshop.

2.2.1 Motion platform

The central item in the Motion-Lab is the Maxcue motion platform from Motionbase (see Fig. 2.6). It was built and designed after the Stewart platform principle: Two bodies are connected by six legs, which can vary in length (Fichter, 1986). One of the bodies is traditionally named base and the other platform. In our case the base is connected to the building letting the platform move by changes in the six cylinder lengths. The Maxcue motion platform has six electrically driven cylinders which are symmetrically arranged between base and the platform frame[9]. The platform is able to perform movements in all six degrees of freedom (DOF), so it can turn around three axes and move in all three linear directions independently. The technical details for maximum displacement, velocity and acceleration are summarized in appendix C. The coordinate system for the platform is identical to the coordinate system for the simulations of the whole lab: The X axis points away in front of the user sitting on the platform, and the Z-axis points upwards which completes the right hand coordinate system with the Y-axis pointing to the left of the user (see Fig. 2.7). Therefore, the rotations around the X-axis is called roll, the one around the Y-axis pitch, and the rotation around the Z-axis is called yaw. Normally those terms are used by pilots, but have been adopted here for the technical descriptions. For describing human orientation in space we use the same names for simplicity[10].

Figure 2.6: Maxcue motion platform

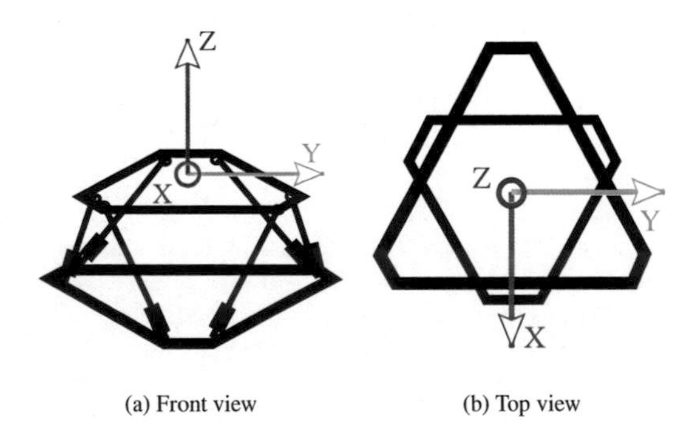

(a) Front view (b) Top view

Figure 2.7: The coordinate system for the motion platform is at the same time the general coordinate system for the simulations for the whole lab.

The actual control of the platform's movements is achieved in several steps (see Fig. 2.8). A data record which contains six numbers, one for each DOF, is given to the platform library at a rate between 30 and 100 Hz. Depending on the filter parameters, these values

[9]Besides a few restrictions on the position of the legs, the endpoints of the cylinders are arbitrary.
[10]Others prefer to use the terms tilt, nick, and heading

can be interpreted as accelerations, velocities or positions[11]. These values are passed on to the DSP board in the motion control host computer by the library provided by Motionbase. This board implements the digital filters in hardware with the given parameters. The filtered values are converted into cylinder lengths by the inverse kinematics of the platform. The cylinder lengths are derived from the given six DOF position by calculating the transformed frame mount points of the legs from the normal setup geometry with one matrix multiplication. The Euclidean distance from the base to the transformed leg positions on the frame is the length of the cylinders. Therefore, there is only one solution for the inverse kinematics. In contrast, the forwards kinematics is more complicated and probably not analytically solvable, but approximately solvable by a multidimensional Newton algorithm for any required accuracy of the calculation. This calculation would, if needed, enable the library to recalculate the actual position of the platform for control reasons given the lengths of the cylinders. Nonetheless, this additional control is not yet implemented in the Motion-Lab Library.

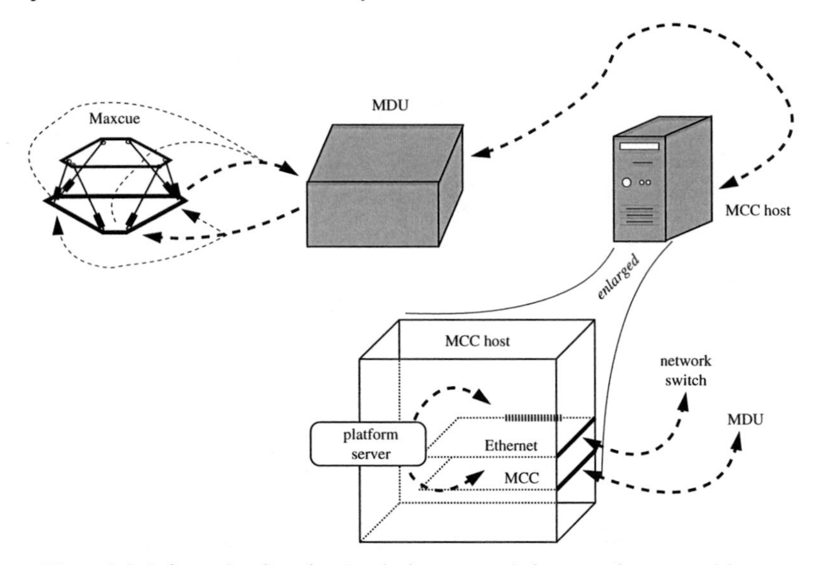

Figure 2.8: Information flow for the platform control: from numbers to positions. The host for the Motion Control Card (MCC) runs the actual server application. This application connects the input and output from the Ethernet with the MCC. On the MCC the platform positions get filtered and transformed into the actuator (leg) lengths necessary to move the platform. Those values are transfered to the Motion Drive Unit (MDU) where they are amplified to control the motors of the platform legs.

There are many similar motion platform systems on the market, mainly being divided into two groups: The legs are either moved by electric motors or by hydraulic pressure. Pneumatic systems can be classified as hydraulic systems, since they share common features. The general advantage of hydraulic systems is the smaller size of the legs; they can generate higher forces with less technical effort. On the other hand, there is always a

[11]The programming of the filters is subject to a nondisclosure agreement and therefore can not be discussed.

compressor needed for generating the force, which is usually very noisy and has to remain close by to allow rapid changes in pressure. The seals of the hydraulic cylinders have the duty of maintaining the pressure inside the cylinder and therefore add high friction to the cylinder. The result is very often a noticeable jump at the beginning of a movement. This can cause disturbances, especially at turning points of one or more cylinders. In contrast, the electric system can start movements more slowly and have smooth turning points. The resolution of the length control for one cylinder can be very high with the smallest step being 0.6 μm in our case. On the other hand, the small steps can cause micro-vibrations, which can be disturbing: The person sitting on the platform can notice the movement by the vibration before it is actually possible to feel the movement by visual, vestibular or proprioceptive cues. In our lab, those vibrations can be covered by very small changes in position driven by white noise which causes constant vibrations and sufficiently covers the actual onset of a larger movement. As the reader can see, each and every system has certain problems with the start of very soft movements. However, the systems also differ in the maximum frequency of movements they can perform. The electric systems are in general faster[12] since the latency of the hydraulic systems to react to small pressure changes is quite high.

2.2.2 Head Mounted Display (HMD)

The visual simulation in the Motion-Lab is presented to the user via an HMD. An HMD combines, in principle, two small displays with an optic lens system, which enables the user to see the small displays at a very close distance while focussing to a comfortable distance. The displays are mounted together with the optic lense system inside a cover. The helmet can be placed on the head of the user like a bike helmet (see Fig. 2.9.c). When considering various models by different manufactures, several points have to be considered. The resolution and technology of the display is naturally important for the visual impression. Recently, LCD's became common and increased the possible resolution. However, because of the illumination decay of LCD, they sometimes present the picture too slowly and afterimages appear. CRT's on the other hand are available in higher resolutions but are considerably heavier. The optic lense system itself is equally important, since distortions and color changes could disturb the presented picture. The most important factor of an HMD is the field of view for the user (Arthur, 2000). Today's best helmets typically cover 40°-60° of horizontal visual field, whereas the human visual field covers more than 190°. The larger the field the more natural the view inside looks[13]. A small visual field, in contrast, can cause simulator sickness and may cause spatial disorientation. Another factor is the weight of the helmet which can cause fatigue. There is usually no external support for the 0.5 to 3.5 kg of an average device. A study by Cobb, Nichols, Ramsey, and Wilson (1999) summarizes the serious effects caused by HMD's as virtual reality-induced symptoms and effects (VRISE).

The helmet we chose to use in the Motion-Lab is the ProView XL50 produced by Kaiser (see Fig. 2.9). The complete technical data are summarized in appendix C. The two LCD's present a visual field of 40°x30° at a resolution of 1024x768 pixels. The refresh rate is fixed to 60 Hz which should be provided by the computer generating the standard XVGA signal. The weight of 980 g is relatively low such that the helmet can be worn for up

[12]Our system is designed to perform active vibrations up to 25 Hz.

[13]Some tasks are also known to require a larger field of view (e.g., driving a very large boat).

(a) Half front view

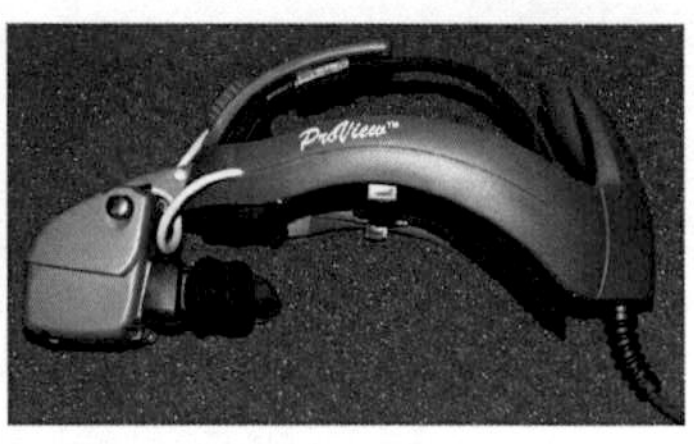

(b) Side view

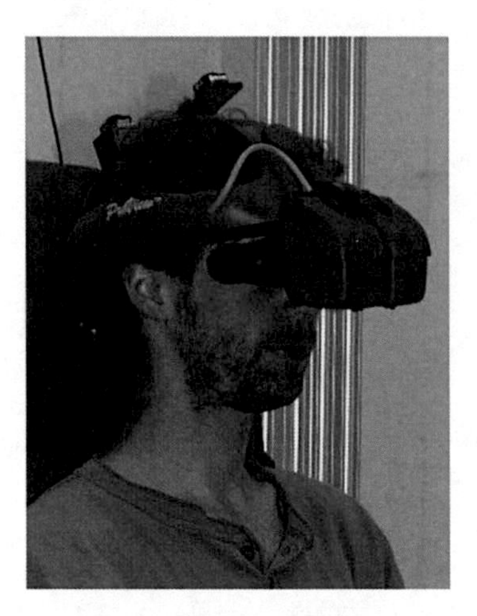

(c) Subject with HMD

Figure 2.9: Kaiser Head Mounted Display ProView XL50

to 60 minutes without discomfort. Compared to projection systems, HMD's have the general advantage that they can easily be replaced by a newer model with better performance/resolution, lower weight and bigger field of view.

Other visualization setups are possible in the Motion-Lab, but have not been implemented yet. In principle, one could use a small LCD projector and a fixed screen both mounted on top of the platform. The projection would have to be carefully coupled with the performed motion of the platform in order to generate a good impression of a stable world. An HMD blanks out all vision of the exterior room, but having plain view around (especially seeing the platform itself) might result in other problems yet to be solved.

2.2.3 Force feedback steering wheel and analog control

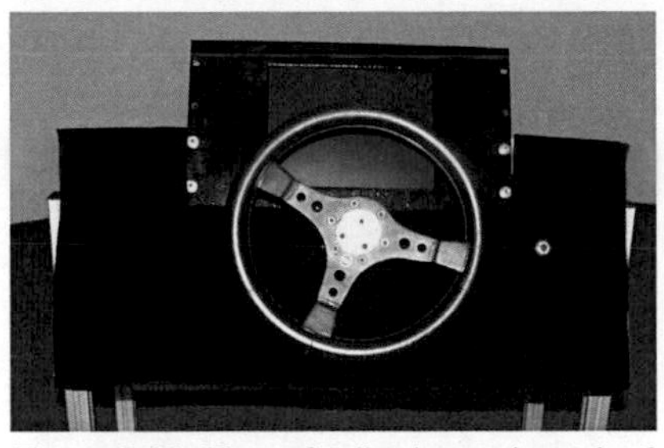

Figure 2.10: Force feedback steering wheel (here in the VE Lab)

The force feedback steering wheel is the primary input device for driving applications (see Fig. 2.10 for the setup in the VE Lab and Fig 2.20 for a picture on the motion platform). This custom built device is constructed to have maximum flexibility and is adjustable for heights, steering angle, and distance to the driver. A high force motor is controlled by an analog card who's signal is amplified and transformed for the motor-control (see Fig. 2.11). A potentiometer measures the current steering angle and enables a fine force control in a local feedback loop. Standard pedals can extend the functionality of the wheel and be used for breaking and acceleration. The current implementation was adopted from game pedals which were connected to the same analog card. The analog card (AT-MIO-10-16E) has more ports which can be used for further extensions. It can sample the data and provide analog output at rates up to 10 kHz. The technical data for the wheel and the analog card is summarized in appendix C. The wheel is also usable in the other lab of the MPI in front of the cylindrical screen (Cunningham et al., 2000a, 2000b).

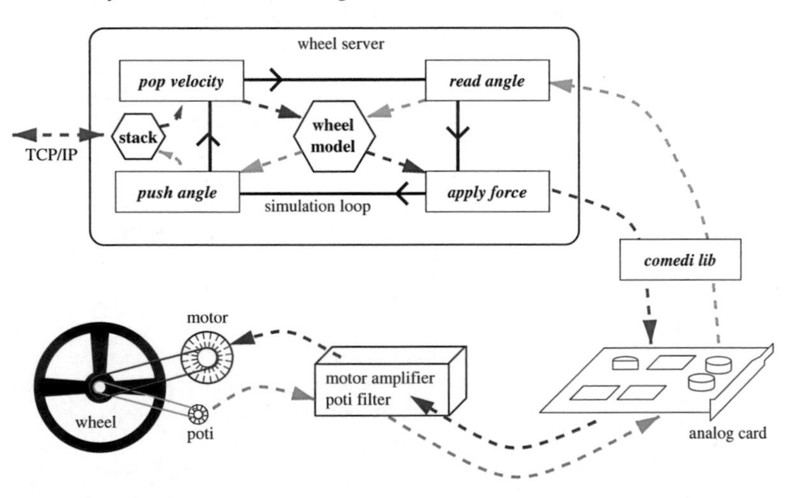

Figure 2.11: The control of the steering wheel: The wheel server gets data via the TCP/IP connection and updates its internal model of the steering wheel. Afterwards the actual angle is read and a force calculated based on the steering parameters. The force is applied to the wheel via an analog card and an amplifier. The steering angle is read out from the potentiometer coupled to the wheel. The angle is converted from analog current to digital values by the analog card. In the simulation loop of the wheel server, this value is transfered back to the stack and sent off to the main simulation.

2.2.4 Joysticks

Joysticks are commonly used in computer games. Therefore, many are constructed to be connected to the standard game-port of a soundcard. Joysticks are handheld devices which enable simultaneous analog control for multiple axes. Simple versions have two axes and more complex joysticks can have up to four axes. Normally, the stick is returned to neutral center position by springs. Modern versions have small motors driving the joystick back enabling the simulation of changing forces. Multiple joysticks can be used in the Motion-Lab (see Fig. 2.12) as input devices to the simulation. The simplest (Fig. 2.12.a) has two separate analog axes and two buttons for digital answers. A more complex device (Fig. 2.12.c) has, in addition, two axes for special controls which should emulate functionality of a helicopter: The foot pedals are negatively coupled and the handle includes up and down movements as well as a digital button emulation for rotations. The last joystick (Fig. 2.12.b) combines the two axes of the first one with a horizontal turn axis and a large number of buttons. This device can give dynamic force feedback if the application addresses a special driver. In general, the control of the joystick is done via the game-port and therefore triggers an interrupt of the system. The maximum data rate is limited in the current implementation to 60 Hz.

2.2.5 Tracker

In order to take the body movements of the user in to account, it is useful to track these movements. This can be done by mechanical devices (like the joystick) or allowing free movements using other tracking devices. Several systems employing different methods are available on the market. Based on high frequency magnetic fields the Fastrak (Polhemus) system or Flock of Birds (Ascension) are the most commonly known. These systems are sensitive to metal in the direct environment and therefore not recommended in the Motion-Lab. Other systems measure the time differences of submitted and received ultrasonic sound signals with multiple microphones (CMS 70P from Zebris) performing triangulation calculations on the data to calculate position. Optical systems provide the best performance, but due to the need for high speed cameras, also have the highest price (Optotrak from Nordern Digital Inc.). See appendix D.5.1 for web references.

For the Motion-Lab we use the IS600-mk2 tracking system from Intersense. This tracking device combines two principles for tracking motions: an ultrasonic system and inertial sensors. Both systems have several advantages which are combined to come up with a six DOF measurement for up to four tracking units. One tracking unit (see Fig 2.13.a) traditionally consists of one inertial cube and two ultrasonic sources (beacons)[14]. The inertial systems are updated at a rate of 400 Hz and therefore provide fast feedback for rotations, but not for linear accelerations. The device has a low acceleration detection threshold under which it cannot record movements. The inertial system is therefore susceptible to slow drifts. The ultrasonic subdevice, on the other hand, works on an absolute scale. Each beacon is triggered by an infrared LED flash and shortly afterwards produces an ultrasonic sound. This sound is recorded by four microphones located at the end of the cross bar mounted on the ceiling (see Fig 2.13.b). As the beacons are triggered in a sequential order, each additional beacon reduces the tracking speed[15]. An individual distance estimate

[14] Other combinations can be configured with a special configuration language via the serial line.

[15] A new version of the system overcame this limitation

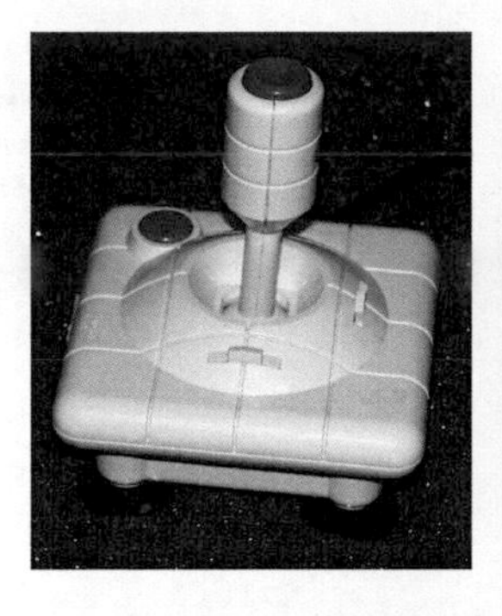

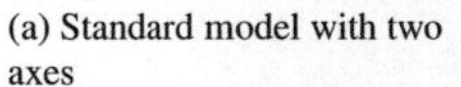

(a) Standard model with two axes

(b) Microsoft Sidewinder with force feedback

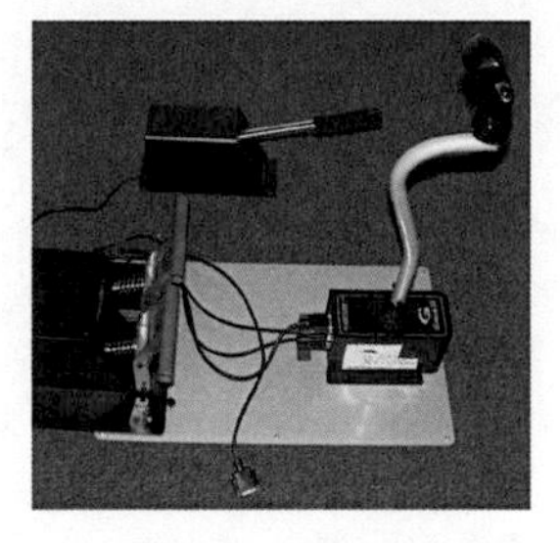

(c) Helicopter control

Figure 2.12: Collection of joysticks which could be connected to one of the game-ports and used for experiments.

is calculated, from the time difference between the infrared trigger and the first sound arriving in each of the four microphones. For each beacon the four values are combined into one measured position in space. The positions of the two beacons of one tacking unit are combined with the data of the inertial cube to yield all six DOF's. The technical data are summarized in appendix C.

Our tracker has two tracking units which can, for example, be used for tracking the movements of the HMD and the platform in all six DOF's. The difference vector between the platform and the subject's head can be used to move a virtual camera for the VR simulation. Naturally, the tracking device can be used for other things as well. For example, one could track pointing movements of an arm or hand.

The communication between the tracking box (see Fig 2.13.c), which integrates the different measurements, and the simulation computer is done via a serial line which causes some additional latency. The overall latency of the system can be improved by separating the translational and the rotational signal. Since rotations cause bigger changes in the rendered picture, it is more important to integrate them with minimal latency. Luckily, the rotations are mostly based on the integrated signal of the inertial cubes which operate

(a) Tracking unit (two beacons + one inertial cube)

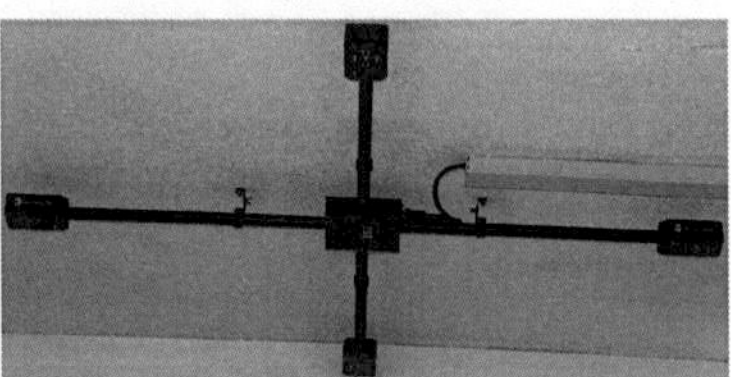

(b) Cross-bar

(c) Communication unit

Figure 2.13: The six DOF tracking device IS600-mk2.

independently of the number of units at a high speed. The rate at which the system is currently used depends on the configuration and lies between 60 and 200 Hz.

2.2.6 Sound cards

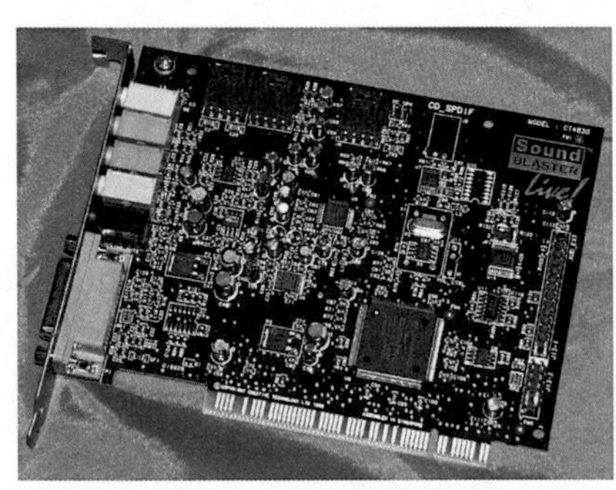

Figure 2.14: The Sound Blaster Live! is used for sound effects, speech synthesis, and for the connection of joysticks.

Sound is generated by a standard sound card (Sound Blaster Live! from Creative) which is shown in Fig. 2.14. The technical data are summarized in appendix C. The Linux driver is currently able to control the sound stream at a rate of 22kHz for both stereo channels. If needed, sound can be sampled in parallel at the same rate. Up to 32 sounds effects or speech outputs can be overlaid at the same time providing a complex auditory scene. We use multiple cards in the Motion-Lab to control speech and other sound effects. Different channels, for example, are used to control vibrations of force transducers (see section 2.2.8). In addition, the sound cards provide the game-port connector for the joysticks (see 2.2.4).

2.2.7 Headphones

As sound is an important feature in VR simulations, it has to be carefully presented to the user. In addition, it is important not to let the user perceive sounds from the real world. Beside the disruption of the immersive feeling, external spatialized sound could provide auditory room context from the real environment. The Aviation Headset HMEC 300 from Sennheiser (see Fig. 2.15) is used to provide sound to the user in the Motion-Lab. These kind of headphones are normally used by helicopter pilots to reduce the noise of the engine. These special active noise cancellation headphones effectively reduce the environmental noise during the simulation, and make the use of external sound sources as spatial references points (auditory landmarks) impossible. High frequency noise is passively reduced by special ear cushions. As the low frequency part of the noise cannot be reduced passively, active noise cancellation is used. The active noise cancellation uses the principle of canceling waves: Fitting the incoming noise, the systems adds the exact same sound with a temporal phase shift of 180° (opposite phase) so that the sound waves cancel out. In addition to the noise cancellation, the headset provides a microphone mounted on a flexible boom. In experiments where a verbal response from the subject is needed, sound can be recorded and provided to the operator. The technical data are summarized in appendix C.

Figure 2.15: Aviation Headset HMEC 300 with high noise reduction

2.2.8 Force transducer

Vibrations are sensed by the human skin. In vehicles, vibrations are often connected to motion. To simulate motion in VR we integrate special vibration devices into the system. In the Motion-Lab, vibrations can either be simulated by the motion platform or by the Virtual Theater 2 (VT2) from RHB which includes the amplifier SAM-200 and two Tactile Transducers FX-80 (see Fig. 2.16). Force transducers function like normal speakers, but without a membrane, transmitting the sound directly to the base plate of the transducer. One can compare them with powerful subwoofers, but force transducers do not generate a sound wave. The motion platform itself can simulate vibrations with high precision in independent six DOF but only up to 25 Hz. The force transducers on the other hand simulate vibrations from about 10 Hz up to 150 Hz. The direction of vibration is perpendicular to the mounting plate of the transducers and therefore only in one direction. Amplifying a normal mono sound source for low frequencies allows the simulation of car vibrations and other "natural" or technical noise realistically. The force transducers can also be used to cover the micro-vibrations from the platform effectively. For detailed technical data, refer to appendix C.

2.2.9 Computer and special graphics

There are several computers in the Motion-Lab with different duties. They are special, either in terms of their OS and library combination or for their special hardware and the

(a) Amplifier SAM-200

(b) Tactile Transducers FX-80

Figure 2.16: The force transducers of the Virtual Theater 2 are used for high frequency vibration simulation.

corresponding library or both. Not all of the computers are used in all experiments, since it depends on the interface and devices used for interactions. For a brief summary of the technical data, see appendix C.

Sprout

This machine is running IRIX 6.5 and recently replaced an older machine running IRIX 6.2. Both OS's are supported with different combinations of `o32`, `n32`, and `n64` library styles. IRIX is used in the lab for the driving dynamics, which are not included in this thesis. Furthermore, IRIX is used in the VE-Lab of the MPI (see 1.1.3) for the Onyx^2 computer displaying VR simulations on the 180° screen. Since the steering wheel could also be used in that lab, IRIX is one of the main clients for the steering wheel devices.

Cantaloupe

This is the Linux computer for the steering wheel control. It moves with the wheel between the two labs mentioned before. The high speed analog/digital card is built in as special equipment for the control of the steering wheel (see section 2.2.3). In addition, the computer can produce sound with the functions from the Motion-Lab sound scripts (see 24, p. 40).

Cucumber

This Linux box is the main computer for most of the simulations. It also connects to the tracking system and to one of the joysticks. It hosts the same sound cards as mentioned before for Cantaloupe. The main simulation is not demanding in terms of calculational power, but in the sense of high reliability of timing: The main loop should run between 10 and 100 Hz depending on the precision and latency one would like to achieve in the simulation.

Borage

This computer hosts the Motion Control Card (MCC) for the motion platform and uses the library provided by Motionbase. In the beginning, this library was only available for Windows95 but was recently extended to WindowsNT. However, we decided to let the system run the old version, since we did not experience any complications[16]. The task is not very demanding, but constant timing for providing new data to the library, and therefore for the platform, has to be guaranteed.

Soy and Tofu

Both machines are identical in most of the technical data and are handled as twins in the Motion-Lab. They run WindowsNT since the graphics driver has the best quality for this OS[17]. The graphics system is designed to provide high resolution images with full screen anti-aliasing in 60 Hz for most of the virtual scenes used in the lab. The graphics power is provided by four graphics cards per machine connected by a specialized board to calculate the anti-aliasing. The graphics system is called Obsidian graphics and is delivered from Quantum3D in combination with the machines, called not without reason Heavy Metal.

2.2.10 Security features

Some security features ensure the safe usage of the lab. It is obligatory to use the seat belt (see Fig. 2.17.a) any time the platform is operated. In the unexpected case of an emergency, the subject can use the emergency switch (see Fig. 2.17.c) to stop the platform at any time during the simulation. The switch is directly connected to the MDU (the amplifier) and activates the security functions. Because of the unknown position of the platform at the moment when the security circuit is interrupted, it is not sufficient to turn off the actuators immediately. The platform could in this case sink to an oblique position due to the mass distribution. Instead, the platform is driven back to the park position (all cylinders short) and after a delay of two seconds after the detection of the emergency halt is physically switched off. The same security circuit can be disrupted by the light beam (see Fig. 2.17.b) which detects persons entering the simulation space around the platform or by the operator himself at the console. More switches can easily be added on demand.

2.2.11 Network and other devices

Beside all the specialized hardware, there are some general devices which are necessary to make the whole setup work. The computers are locally connected in the lab via a high speed switch (3com SuperStack II Switch 3300 with 12 switched 10/100BASE-TX ports). The switch is integrated into the campus net via fiber optics connections. Therefore, the equipment is theoretically usable from all over the internal net given the above mentioned OS platforms and the Motion-Lab Library (see section 2.3.6).

To handle several computers with one set of monitor, mouse, and keyboard, we use a PolyCon console switch. Due to limited space and simplicity reasons, this solution was

[16]The system had once an uptime of more than 180 days!

[17]They also could run Linux, but the support from the manufacturer has its emphasis on WindowsNT.

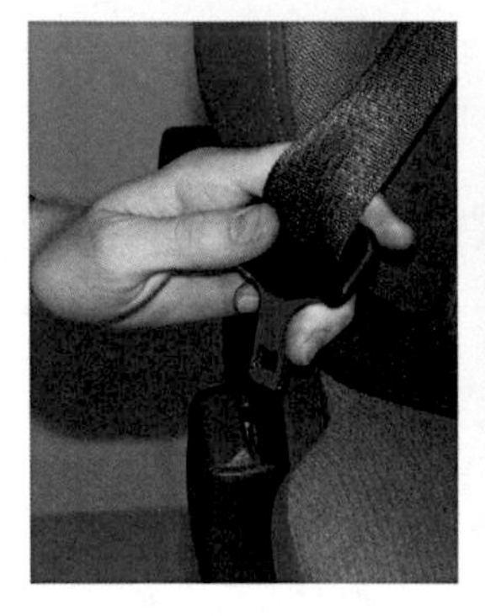

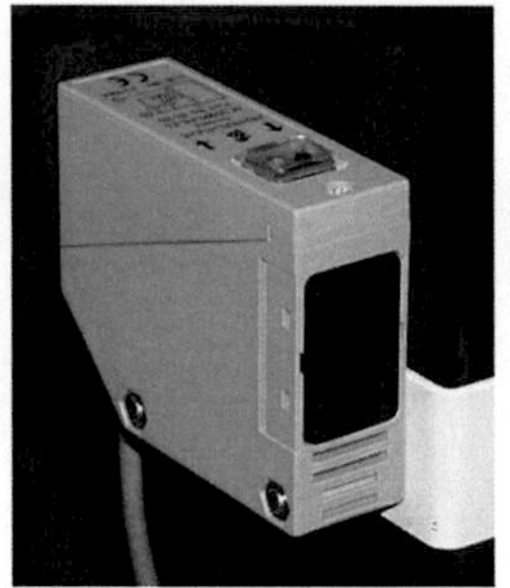

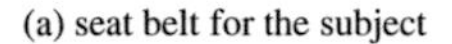

(a) seat belt for the subject

(b) light beam at the door makes sure no one enters the danger zone

(c) emergency break for the subject on the platform

Figure 2.17: Security features are used in the Motion-Lab to ensure a secure usage of the lab. The subject has to use the seat belt which is built into the seat (a). In case of an emergency, the subject can stop the platform at any moment without the help of an operator with the emergency break (c). The operator has a similar switch to stop the platform. The light beam (b) is stopping the platform as soon as a person is entering the close space around the platform.

chosen instead of a whole range of monitors and keyboards for controlling the different computers. Most of the current work could be done via the network, but especially for graphics and quick status overview it is more comfortable to have the actual console close by. The computers in the lab are secured against power failure up to several minutes by means of a Smart-UPS 2200. This device uses batteries to provide 220V power even when the power network fails. For safety reasons it is important to have the control computers running, even when the platform itself stops. The behavior of the motion platform during a simulation where just the control PCs fails would be unpredictable.

2.3 Software

In a normal thesis it might look awkward to talk too much about the software which was used to achieve certain goals. The software is often seen as additional requisite because it did not take long to develop it. Very often it does not fit into the thesis because of the general topic. Surely, a lot of programs just help us to do things and they stay in background as tools one does not talk about.

On the other hand, a thesis can introduce general software concepts. They can either stand for themselves or – even better – be realized and thereby proven to work. In this case, it makes sense to talk about the concept and the realization together. A mathematician who shows in a complex proof that a solution exists might be satisfied, but others will be pleased if he demonstrates one solution. Taking computer science seriously, general ideas and their realization should go together as proof of function.

This section of the Motion-Lab chapter therefore presents the basic principles of the implementation of the Motion-Lab Library. A certain style of software development guided the implementation and concepts which will be introduced. Last but not least, this section documents the border between programs and libraries developed for the lab and components written by others but used for the realization. This project would not have been successful without a lot of different tools and libraries.

2.3.1 General software environment

The commercial operating systems (OS) used are Windows95, WindowsNT, IRIX 6.2, and IRIX 6.5, and were bought from the respective companies (Microsoft or Silicon Graphics). The low-level library (for Windows95) for controlling the platform was included in the delivery of the Maxcue motion platform from Motionbase. For the development of the simple 3D model of the second experiment (see chapter 4) the program Multigen was used. The rendering of the model on the specialized graphics systems (see section 2.2.9) involved the usage of several commercial 3D graphics libraries. At the moment, the Motion-Lab Library supports rendering with Performer from Silicon Graphics for IRIX, Vega which is distributed by Paradigm, and OpenGVS, a product of Quantum3D. Vega and OpenGVS are running mainly on WindowsNT, but IRIX libraries are also available.

The other software parts of the system (open source or freeware) are freely available on the Internet at the address given in appendix D. Mainly, the Debian GNU/Linux Distribution and ACE as a general base for the software development have to be mentioned. The Performer graphics library is also free for use on Linux. Since the freely available parts provided the base of this project, the author will consider making the Motion-Lab Library public under GNU Library General Public License with appearance of this thesis.

The Motion-Lab Library was designed and mainly implemented by the author. Under his supervision, Tobias Breuer helped with the implementation of smaller parts as documented in the source code. Documentation (apart from this thesis) is done in DOC++ which provides HTML and LaTeXversions of the C/C++ structures and class descriptions.

Before we go into the details of Motion-Lab software, the main tools and packages are introduced to the reader in case they are not known. Without some of these packages and programs, the development would have taken much more time and it might have been impossible for the author alone to complete the system and reach this high level of abstraction and perfection in the interfaces within just two years.

ACE - Application Communication Environment

Without ACE, most programming for different OS is much more difficult, error prone, and tiresome. System calls differ across OS in details and interface. When it comes to multi-threaded programming and socket communication, the programmer is forced to learn a lot of small differences for each and every OS. ACE, on the other hand, provides one single interface for most system calls, multi-threaded programming and socket communication. One has to learn only the specialties of ACE instead of those of 4-5 different OS. Since the realization is mostly done with inline statements[18], saving all the costs for additional

[18]Of course, only in those OS and compiler combinations which allow those statements.

function calls, ACE does not add significant overhead when the programs are run. ACE simplifies the realization of software for multiple OS concerning plain C++ code. When it comes to direct access to hardware, such as serial ports or even more special things like analog cards, ACE admittedly does not help any further.

sox and all the other well-sounding names

The sound is realized by a collection of ten small programs, each doing part of the job and solving one small problem. The main part for the actual replay is `sox` which works together with the kernel sound module. The program converts different sound formats into the format which can be played by the low level sound driver implemented in the kernel module. The script `play` wraps the more complicated and tiresome options of `sox` and makes the handling easier. For the replay of sound files, two more helpers allow playing a loop without sudden breaks in the sound stream. The buffering is done by `bag` and the repetitions are controlled by `repeat`. For the other application, speech synthesis, more scripts are required. Four filters modify the letter stream by speaking '@' as 'at' (done with `sed`), removing line breaks with `pipefilt`, replacing numbers with spelled-out numbers (realized with `numfilt`), and subst. abbr. w| the l. vers. they st. 4[19] (included in `preproc`). The actual translation of text with the corresponding phonemes is done in `txt2pho` so these can be pronounced by `mbrola`. Multiple speakers are available in the database for the pronunciation which allow, in addition, the usage of speed and mean frequency as independent parameters. In the end, the combination of `play` and `sox` plays the sounds for the given low-level sound driver. Both functionalities were summarized in two scripts by Michael Renner with some help of the author. For the Internet addresses of the different scripts and tools see appendix D.

CVS - Concurrent Versions System

The CVS, like other version control systems, is a tool that provides a database which keeps old versions of files along with a log of all changes. It operates on the hierarchical structure of the file system containing version controlled files and directories. It enables multiple authors to edit the same files at the same time and tries to resolve conflicts, if possible. The single copy of the master source contains all information to permit the extraction of previous versions of the files at any time either by name of a symbolic revision tag or by a date in the past. There are versions of CVS available for all OS used in the Motion-Lab. All files (source, configuration, models, and documentation) are managed with CVS. Working with one master source code enables the Motion-Lab users to share latest versions and avoids keeping bugs in different versions of the same functions. Commonly used functionality therefore becomes more stable and powerful over time.

DOC++

The DOC++ documentation system generates both LaTeX output for high quality printouts as well as HTML output for comfortable online browsing of the source code. The program directly extracts the documentation from the C/C++ header file or Java class

[19] substituting abbreviations with the long version they stand for

files. For the Motion-Lab, additional files are included in the documentation for the description of devices. The source code containing the DOC++ tags provides two versions of documentation: The complete developers documentation with all inside functions of the library and a user version as a subset that concentrates on the parts which are "seen" – that means usable – from the outside.

GNU-tools: gcc, gmake, emacs, and others

The famous collection of GNU-tools is the base of every Linux system. Furthermore, most of the tools are available for IRIX and some even for Windows. The compilers gcc and g++ provide useful hints when struggling with errors. Without gmake, the ACE package would not be able to compile easily. Finally, all the source code, documentation, and this thesis were written with the help of emacs. The useful aids and tools help a lot and make programming in an UNIX environment enjoyable.

2.3.2 Distributed programming for multiple OS

The above section gave an overview of which software was used in the development of the Motion-Lab Library and which software is still running hidden inside some scripts. However, having great libraries like ACE and powerful tools like CVS is not the complete story of how distributed programming for multiple OS becomes successful. In this section, some general guidelines introduce a more general concept than "take this and that and it will work".

What is the general flow of information in the project?

This question leads back to an old principle of software design in which one has to draw a lot of boxes with the information flowing between those boxes. Those diagrams help us to get the big picture of a process or program. Main sources and destinations of information need to be identified and put into the framework of informational flow. Surprisingly, programmers rarely do those drawings – but why?

Small projects tend to work out after a couple of attempts and redesign stages. Small projects grow slowly enough to forget about structure on the way, but tend to grow too fast to let real structure build up. Small projects start all over again, after reaching the point where "small" additions become more and more the purpose of a whole new project. The general mismanagement of all three assumed scenarios is the missing general flow of information, which should guide the project from the beginning. Therefore, one has to start with the question:

Where does the information come from and where should it go to?

If at least the points where information comes from and where it should go to are known, the parts in between very often become trivial. It is reduced to the question of conversion, recalculation, and managing information. Sometimes, it looks like the information is necessary at all points at once. In a distributed network, but also in a single computer, this leads to enormous data transfer which will slow the system down. The analysis should focus on the question of whether the complete set of data is necessary or if there are points where a smaller part would be sufficient. Following this question, one could come up with a structure of necessary data pools which might divide the project into different units/parts. This structure might be different from the one obtained from analysis of the source and destination of information.

How do special OS/library requirements split up the project?

Yet another structure might become clearer by looking into the need for special libraries or hardware which are involved in the project. As mentioned in the hardware section of this chapter, special devices often come with special libraries. Since most of those libraries are not binary compatible, they have to be used on the given OS platform or re-implemented on a different one. The re-implementation very often is made impossible by the manufacturer for reasons of fear: They fear the loss of knowledge and their position on the market by making interfaces documented and open to public[20]. Even if the interface is well documented, it will take a while to re-implement everything from scratch. Therefore, it is generally easier to take a look into the given libraries and design the structure of the project in parts around those libraries. The client-server concept (see section 2.3.3) provides an additional layer of abstraction.

Is it necessary to run the parts under several OS and on different computers?

One is lucky, if one can answer this question with a "*no*". On the other hand: Use the opportunity to stay flexible and gain the advantages of independence! A system which is based on parts that do not care about system requirements is more likely to be used by a lot of people. Not only the power of multiple users working on the same project, but also the power of parallelly working machines should be persuasive. The disadvantage of missing load-balancing in a situation where one computer does one job could be overcome by having several computers doing similar jobs and distributing the work load among them. Being able to use multiple computers for the same job, decreases the probability that none of the computers work (see section 1.2.1).

Combine all the above structures to come up with a plan!

In the process of planing a new project that involves different hardware and software requirements, the above thoughts might help one to come up with different solutions for the future structure. Depending on the importance of different requirements for a given project, one has to combine the results from the above questions. It could actually help to design the project in multiple ways and join those parts into a final concept which share the common structure of multiple solutions. With this approach one can almost be sure not to miss an important part which becomes obvious once the implementation has been started.

[20]As the development of graphics support in the Linux community has shown, the manufacturers might very well profit from the community by giving the interface to ambitious programmers and participate in distribution of their products.

2.3.3 Client - server architecture

The communication, library, and hardware structure of the equipment in the Motion-Lab suggested a client - server architecture based on the following requirements. Some of the libraries were explicitly designed for one OS and therefore, at least three different OS had to be used (Maxcue low-level access on Windows95, graphics like Vega/OpenGVS on WindowsNT and the driving dynamics[21] on IRIX). The control proposed by Motionbase for the Maxcue motion platform is based on UDP broadcast calls, which were unacceptable in our local network for two reasons. First and most important, the protocol had no security check and anybody could simply add data to the stream by opening a telnet-like program on the specific port with the result of unpredictable behavior of the platform. The other reason was to reduce overall traffic and not to disturb others by using the platform inside our local network. In the process of solving the security issues, it became clear that a constantly running server controlling the platform and accepting only authorized connections, one connection at a time, would enforce a more deterministic behavior and control for security risks. Similar requirements had to be considered for the control of the steering wheel, since subjects in the experiments directly interact with the device and should not face any kind of risk. Even in cases were the VR simulation fails due to errors in the program, the server can still run and guarantee a safe shutdown of the device.

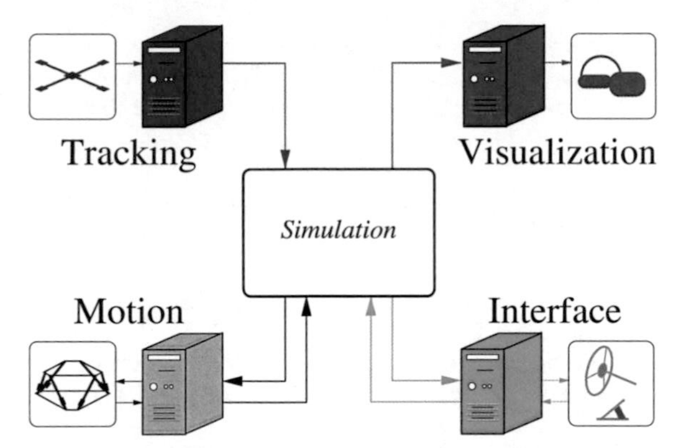

Figure 2.18: The general client - server framework for a distributed VR simulation.

The idea of server programs controlling the different input and output devices is powerful not only for security issues. It also forces the implementation of a strict layer of abstraction between a client who wants to have some data from a device or send it towards the device, and the controlling server that implements the specific input and output behavior. Having this abstraction layer established not only for the platform and steering wheel, but also for all the other devices like joystick, tracker and visual display, the simulation logic becomes quite independent of special hardware and the actual physical realization. It is now possible to exchange the graphical rendering without affecting the simulation of a virtual observer, which is quite unusual for most VR systems. The advantage of this became clear when the actual machines were exchanged and we had to move from the

[21]The driving dynamics done by Müller System Technik is not part of this thesis, but had to be considered in the design.

Vega libraries to the OpenGVS rendering. The main simulation was not subject to any changes, only the graphical rendering had to be re-implemented based on the new library. This example illustrates the flexibility of this approach enabling the integration of new hardware, which very often comes with specialized libraries. The server implements the abstract device layer, based on the interface of the specific physical realization, and the clients just stay the same. This enables multiple users to program for the abstract interface rather than for a specific device, which enhances the overall usability of software. Even changes in the OS or the physical device do not affect the VR simulation.

It also became easier to add new services for all simulations. Since the main part, the direct handling of the devices, is outside the actual VR simulation of the user, the changes in the simulation program necessary in order to use additional services are reduced to a minimum[22]. Taking the client server concept together with asynchronous communication, it is easy to have devices work at different update rates. Each device can work at its necessary speed and provide either local feedback (like the steering wheel) or global feedback (like the visual display reacting on a turn of the head). The information flow is depicted in Fig. 2.18 for a simple VR simulation based on the client - server concept.

2.3.4 Use of templates

Having introduced the client - server architecture, it becomes obvious where one should use different classes and which functionality should be shared between different streams of information. In order to make the communication between the client and the server as simple and efficient as possible, one step for confining the communication was taken. The data packages themselves should contain only similar things in sense of data type lengths. The allowed types for the data inside data packages were fixed to `int` and `floats`, both containing four bytes and sharing the same conversion for changes between processor types[23].

All those points do not connect to template discussions at first glance. Templates in C++ enable programming for unknown types to some degree. The concept is nearly as powerful as libraries are for sharing functionality for known types. It allows implementation of general functions in a type safe way at compile time[24]. In combination with inheritance, it becomes powerful, as the reader can see in section 2.3.6 where the realization of the different devices clients is discussed. Another example of efficient usage of templates can be seen in the stack implementation[25].

Connecting both ideas, it became fast and easy to send data packages over a TCP/IP connection with the above restrictions. General template functions were used to pack and unpack data packages and store the data on stacks. Since all data could be handled in the same way, there were no functions necessary for extracting special things like pointers of strings. Avoiding dynamic memory calls made the implementation efficient and safe to memory leaks. Extending the known data types with additional variables is easy since only initialization and print routines have to be changed due to their additional text output naming the data; the data communication routines stay untouched.

[22] For example, it was possible to add head tracking into a simulation by adding less than 20 lines of C++ code into a normal simulation.

[23] see differences of big and small endians in the SGI ABI or the ntoh* man pages.

[24] This is not true for void pointer concepts, which test class membership during runtime.

[25] The stack implementation is special for not keeping everything on the stack, as the concept normally suggests.

2.3.5 What is real-time?

Defining "real-time" is highly dependent on purpose and context. "Only reality runs at real-time" could be one statement, which could be opposed by some consideration about the brain of a bee and a CRAY high end computer with eight processors. Both the bee and the computer have roughly 10^6 neurons or transistors respectively. For the bee one "operation" is quite slow with $10^{-3}s$ compared to $6 * 10^{-9}s$ for the computer. Nonetheless, due to the high parallel execution in the bee's brain, the bee theoretically executes 10^{12} operations per second while the CRAY stays behind with only 10^{10} operations per second. What is real time for the bee and the CRAY? The bee behaves in the world with a considerably high latency, particularly if one compares it to the possible high precision of the CRAY. However, the speed of the bee's reaction is sufficient to do navigation, pattern recognition, social interaction and more specialized things. The CRAY computer is faster for very specific and simplified tasks but could not reach the complexity of the bee's perception and interaction with the world. This consideration holds for humans as well as for the bee. Looking into different modalities, the latency and time accuracy of perception differs widely. The time lag which would be acceptable, that means which would not be noticeable as additional offset to the "true" point in time, is different from the overall time resolution in each modality. Programming a VR setup, where things should behave "normally" and allow us to perform without previous training, has to consider both time constrains: The update rate and the latency. The goal is therefore not to define real-time in yet another way, but to provide sufficient fast interaction in the sense of update rate and latency.

A number of studies have shown either the neural latencies of the human system or the necessary update rate which should be provided by a simulation. For example, Dell'Osso and Daroff (1990) refer to several interactions between the vestibular and visual system. The latency for head movements which result in the vestibulo-ocular reflex appears to be less than 15 ms. The same study specifies that the latency for eye stabilization control is larger than 100 ms for field motion and more than 125 ms for corrective saccades for positional errors. The latency for extraction of optic flow seams to be greater than 300 ms (van den Berg, 1999). The eye stabilization mechanisms therefore react to changes in the vestibular system much quicker than to the perceived visual stimulus. The effect of apparent motion is visible for image changes faster than 18 Hz. Normally, update rates of 25 to 30 Hz are used for computer rendered pictures. The accuracy of audio localization also depends on the frequency of the provided stimulus. King and Oldfield (1997) described the necessary spectrum of the sound stimulus: "Results show that broadband signals encompassing frequencies from 0 to (at least) 13 kHz are required in order for listeners to accurately localize signals actually presented from a range of spatial locations". The human skin is known to be sensitive for vibrations faster than 2 kHz. For vestibular stimulation, update rates of 250 Hz were used to create a smooth path (Berthoz et al., 1995). VR simulations have to match these numbers, and provide sufficiently fast update rates, and react to changes like the turn of the head with minimal latency. In addition, it is important to keep the time offset between stimuli presented to different modalities sufficiently small in order to not disturb the natural integration process.

2.3.6 Motion-Lab Library

The realization of the above concepts in the frame of a thesis will stay away from printing pages and pages of source code. The general implementational details are given without referring to actual code. The code, the actual data structures, and the respective documentation can be found at the address given in appendix D. All the actual C++ code compiles on all five OS platforms involved, as long as it does not concern special hardware libraries.

The overall structure of the Motion-Lab Library is simple. The library provides clients classes for all devices in the Motion-Lab. Those clients control the communication to the device servers, making the communication transparent to the user. In addition, some useful tools and functions for matrix and vector handling, and keyboard control are provided, but not discussed here in detail.

Stacks

Generally a stack is considered to be a structure which keeps information in a "last in, first out" (LIFO) fashion. The stack for the Motion-Lab communication works exactly with this principle, but with one unusual addition to it. There is only one "last" for output available and packages pushed earlier will be lost. In exchange, the stacks provide information about how often the last record had actually been popped. Of course, a template class is used for the realization of this concept, since the data are not touched at any point in the stack. In addition, the implementation is thread-safe for multiple readers and one writer without locking write or read access. Writing processes add new information and simultaneous reading processes are guaranteed to get the most recent data available. Internally, the stack class uses a ring structure for the storage of information.

This behavior is required for one simple reason. The asynchronous communication explicitly demands high speeds which may include dropping of whole packages. Locking would reduce efficiency, and always providing the newest package reduces latencies in the system. A queue concept would force the accepting process in a communication to check whether there is more information waiting to be processed. The stack concept on the contrary provides the most recent information available and drops other packages without the risk of increasing queue length.

Communication & devices

The communication itself is hidden (as computer scientists like to say, "transparent") to the user. It involves a collection of templates realizing the *send* and *receive* as *client* and *server* via a TCP/IP socket. Gathered as a virtual device, those templates implement the transparent communication with one specific server. As a result, the actual implementation of the abstract device for the user is done in the library by defining the data structure for the communication for one type of client. Therefore, the user does not have to worry about the actual template usage or class instantiation for the communication. The user is addressing a device by instantiation of the specific client for this device. Naturally, the server for that device has to run beforehand[26].

[26]Since those servers can run all the time, one should consider running those as daemons starting at normal system startup.

Sound scripts & functions

There are two simple scripts enabling the use of sound in programs running on Linux machines in the Motion-Lab. The corresponding functions for C++ programs are available in the Motion-Lab Library. There is not yet a sound server which could be used for programs running on different machines and OS. In general, the sound is realized by a collection of small tools (see page 40) which are available for free on the Internet.

- `ml_play`: Replays all well known sound formats either as an infinite loop or a given number of times. Internally, the sound is buffered for the loop to guarantee break free replay even under high load of the computer.
- `ml_say`: Speaks a given string with German pronunciation. The string could include numbers and abbreviations. The speech synthesis is done online, so status messages or other feedback could be given to the user.

Device servers

At the moment, there are several device servers available, some running on multiple OS platforms, but most of them are confined to a specific one due to hardware and library constellations. In general, it makes sense to run the server on one specific machine, since the server has to connect to some physical device. Different ways of connecting the actual device are handled inside those servers. Additional servers can easily be implemented by using the existing servers as examples.

- platform:
  The control for the vestibular simulation runs on Windows95 and connects to an analog card with specialized library from the manufacturer.
- hmdvega, hmdperformer, and hmdgvs:
  The visual simulation runs on WindowsNT, Linux or IRIX and displays the rendered 3D model via the normal graphics output. On WindowsNT and Linux the output of the graphics card can be connected to the HMD. Stereo vision is possible by using two machines which synchronize their output.
- wheel:
  The force feedback steering wheel control runs on Linux with the help of an analog card and a library from the LinuxLabProject.
- joystick:
  It is currently possible for Linux to control up to two joysticks at a time via the game-port of one sound card.
- tracker:
  The tracker server reads out the serial port of a Linux machine and interprets the binary or ASCII output of the Intersense tracker.

2.3.7 Applications

Several applications run in the Motion-Lab with different goals. A simple driving demo will be presented after explaining the concept of the virtual observer for VR simulations based on the client - server architecture. In addition, two experimental programs are outlined to give an impression of how things might look for open- and closed-loop programs. Nonetheless, the general structure of all programs is nearly identical since the tasks are quite similar.

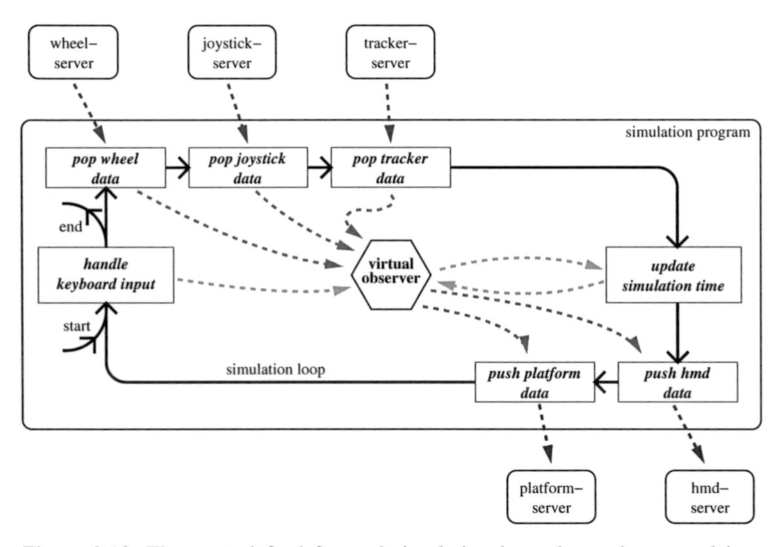

Figure 2.19: The general feed forward simulation loop shows the general input in the upper left-hand corner and the output in the lower right-hand corner. The virtual observer in the middle gets influenced by the input from steering wheel or joystick and the head tracker. Based on the internal representation of the observer's movement status output data is generated for the platform and visualization.

The overall structure works around a central simulation loop as depicted in Fig. 2.19. The input devices provide information which is used to update the status of the virtual observer. Based on a time difference Δt in the main simulation loop the observer's movement status is updated describing a discrete version of what happens in the real world (velocity: $V_{t+1} = V_t + \Delta t * A_t$ and position: $P_{t+1} = P_t + \Delta t * V_{t+1}$). Naturally, one could add friction, wind resistance, surface slant and other factors here to slow down or accelerate the observer, based on the model of the world. For simplicity, those factors are assumed to slow down the observer (like sliding on a flat horizontal plain) a little bit and change, therefore, the velocity by a damping factor. After having updated the observer's internal status, the output devices get new data from the simulation. At that point, we can have a short time delay before we start all over again, in order to control the speed/update rate of this simulation loop.

Simple driving demo

The driving demo incorporats joystick and/or steering wheel control. Therefore, most of the above described equipment is actually involved in this simulation (see Fig 2.20). The car dynamics are kept as simple as possible, since it should soon be replaced by the professional one as mentioned earlier in this chapter. However, it is just a demo and not used for experiments in its current state. The general structure is exactly as described above. The virtual observer gets input from the different input devices and sends commands to the visualization (HMD server) and the vestibular simulation (platform server). The data for the visual simulation are the actual position and velocity of the virtual observer. In contrast, the data sent to the platform could not be transformed into a movement of the actual distances in meters, but had to be reduced or scaled (see "motion cueing" in section 2.1). One simple solution is to send the actual speed of the observer as positional data for forward movements and negative pitch. If the observer gains speed, the platform pitches backwards to substitute for some of the linear forward accelerations. As a cue for higher velocity, vibrations are simulated with increasing amplitude. The combination of both results in a quite realistic feeling for driving without sudden changes in velocity. Since the platform can only rotate a certain angle, simulated turns have to incorporate the same principle. Adding some roll motion to simulate tangential forces enhances the realistic feeling. Nonetheless, psychophysical tests have to be performed to match data from real drives to simulated ones and to equalize both at the perceptual level.

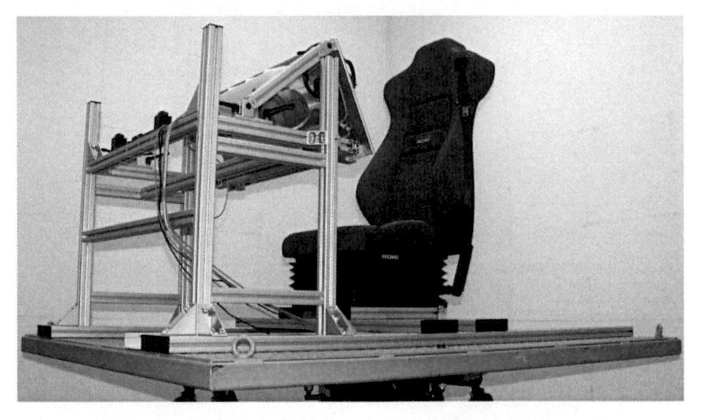

Figure 2.20: Driving setup with force feedback steering wheel.

Experiment 1: Distance, velocity and acceleration judgements

Before we can start to program realistic driving simulations on the motion platform, we have to know which parameters are actually perceived by humans sitting on the platform. Therefore, an experiment was designed focusing on the question of perceptual parameters (see chapter 3 for detailed experimental design). Concentrating here on the software issues, it should be mentioned that this experiment was explicitly done with an open loop paradigm. The subjects had no influence on the performed movement, but had to report verbally on their perception of distance, velocity and acceleration. The simulation was, therefore, independent of user's input. The movements were predefined with Gaussian shaped velocity profiles by a parameter setting the controlling distance and the maximum acceleration. The positions for the movement were calculated beforehand and played back with the appropriate timing. This program shows that the client server concept can also be used in feed forward (open loop) conditions where no interaction is required.

Experiment 2: Holding balance, and coding vestibular and visual heading changes

The second experiment involves, in contrast to the first one, continuous feedback from the subject. It therefore uses a closed loop paradigm. All the details are described in chapter 4. Concerning the program itself, the structure is in principle quite similar to the driving demo. Because of the experimental conditions and different stages during the experiment, the conditions had to be scheduled based on the performance of the subject. Parameter files describe thresholds and dynamic changes in the level of difficulty for the task. The task itself was to stabilize the platform for roll movements based on vestibular cues. The visual simulation was not providing any roll information. The disturbances increase in speed and amplitude for higher levels of difficulty. In the end, the subjects had actually performed heading changes based on the paths they had learned. The relationship between visual and vestibular heading change was controlled and changed only in the test condition. Therefore, it became possible to ask whether the vestibular or the visual turns where encoded in the learning stage. Changing those relationships is easily achieved in VR and would otherwise be very difficult to perform.

2.4 Summary of system characterization

In sum, the Motion-Lab implements a VR setup that enables psychophysical experiments with a variety of hardware in order to simulate multiple sensory inputs. The complex simulation is distributed across a network of specialized computers enabling input and output for intense interaction. Different experimental paradigms can easily be implemented with the Motion-Lab Library which effectively hides from the programmer problems that are imposed by the distributed systems approach.

The lab uses a network of standard PCs extended by special VR equipment. The different units of the networks are exchangeable and share multiple resources. The system can therefore be classified as distributed system. The general architecture realizes a client/server approach in which each hardware device is managed by a specialized server. The servers implement abstract devices, which efficiently hides the differences of various connected hardware. The multi-threaded library provides the clients for the abstract device interfaces. Therefore, these clients connect to the servers, hidden from the VR application programmers view. The bidirectional communication is asynchronous and done via TCP/IP with low latency. Specialized stacks circumvent problems with different frequencies which are imposed on the system by the demands of different sensory modalities. Smoothness of the data stream is established where needed by inter- or extrapolation methods. The library is available and used for multiple OS, hiding again OS specific interface differences of Windows95/NT, IRIX and Linux. Distributed development techniques were used for concurrent access by multiple programmers.

VR simulations are bound to include multiple senses. The goal typically is an immersive simulation which allows the user to interact with the environment with acceptable latency and high degree of realism. The observer's movements are usually unnaturally done in

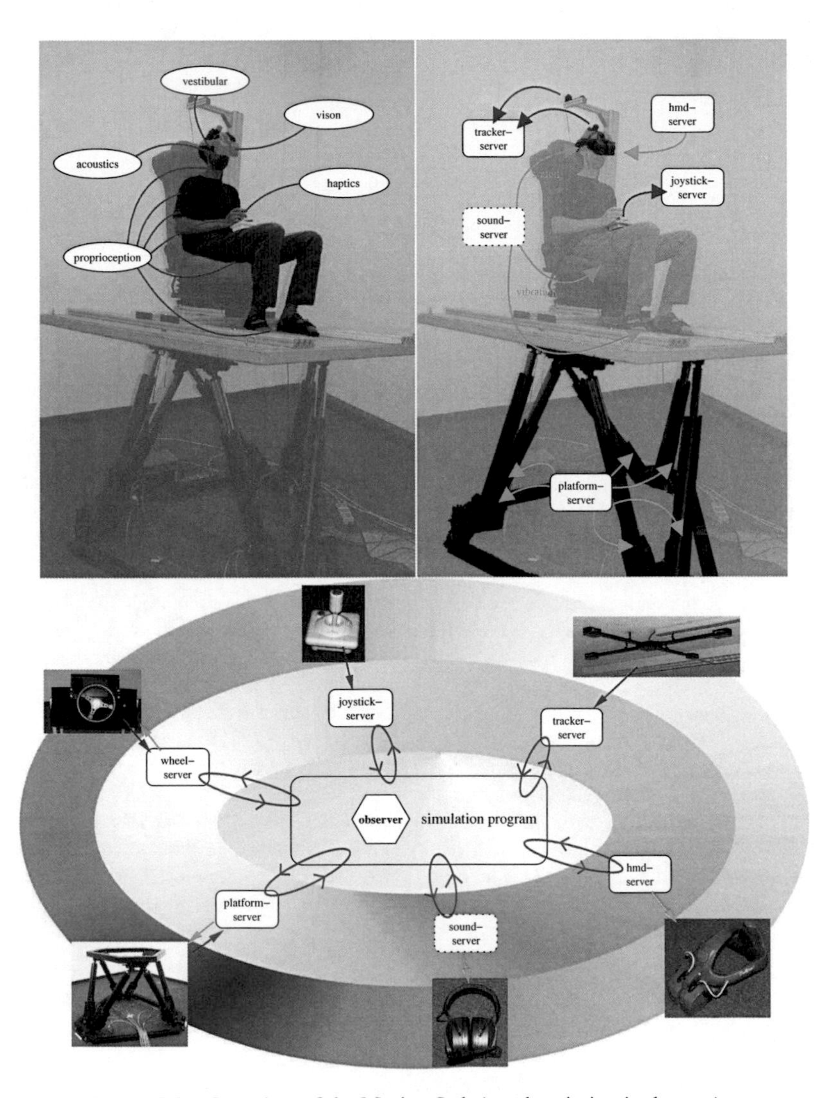

Figure 2.21: Overview of the Motion-Lab (see description in the text).

the simulation. Due to the lack of spatial updating, users do not feel that they are moving in space. In contrast, the present lab realizes realistic and immersive simulations for multiple senses. The latency between an action of a user and the feedback of the system is reduced by multiple layers of feedback loops within and between modalities. Specifically, vestibular, visual, acoustic, vibration, and haptic stimuli can jointly be used by the applications:

- The **vestibular** stimulation is realized by a six degree of freedom (DOF) motion platform with can perform high accelerations;
- Stereo **vision** is presented via a head mounted display with high resolution realized by different graphic libraries rendering the virtual scenery;
- **Acoustic** simulation is not yet presented in 3D, but already includes synthetic speech generation for multiple speakers and numerous stereo sound effects;
- The sound generation can be employed for low frequencies **vibration** stimulation;
- The simulation of **haptic** force feedback for the steering wheel delivers a realistic experience for driving simulations.

Other input devices can be used for typical VR interaction. Joysticks allow analog multi-axes control of quantities coupled to immediate changes in the virtual environment. Trackers for six DOF are used for the input of pointing movements as well as head tracking for the control of the virtual camera.

Figure 2.21 summarizes the client server architecture of the VR simulation in the Motion-Lab. The top part of the diagram shows the equipment of the lab as well as the senses of the human observer that are involved. The lower part depicts the implementing architecture on three levels: The inner software level of the VR simulation of the virtual observer, the outer hardware level of the VR equipment and the level in between formed by the distributed device servers implementing the abstract layer of I/O devices. The interaction of observer and virtual observer is realized by multiple feedback loops connecting different levels. The human observer perceives the virtual environment through multiple senses and interacts with the simulation via tracking, joysticks and the steering wheel. The virtual observer sends and receives data from the devices and simulates the VR environment as a discrete version of world. In sum, the Motion-Lab is usable for closed and open-loop experiments. Therefore, various psychophysical experiments in VR were made possible by this distributed system.

Chapter 3

Experiment 1: Distance, velocity and acceleration judgements

Under normal conditions, our spatial location (position and orientation) in the environment is known to us and is self-evident. In typical virtual reality applications, the relationship between (mostly) visually perceived location and the real body location is broken. Consequently, our position and movements in space are no longer coherently perceived by all our senses. The visual sense is not the only one that provides a spatial frame of reference. We also perceive spatial auditory cues and form reference frames of our own body position and motion by proprioception, and external accelerations by the vestibular system. If those senses agree on one reference frame, we feel in a stable environment immersed. There is hardly any way to misperceive one's own location, which might be the reason why humans have difficulty ignoring the stable world. Moreover, some people believe this basic assumption of a mostly stable world is enhancing our perception because it can reduce processing effort. The assumption of the stable world around us reduces most of the analysis to the changing parts. However, people argue, there must be some sort of boot strapping process which allows us to reset or correct our spatial frame of reference in the environment. Of course, it is possible that both principles exist in parallel. The faster update process would analyze the changing parts and a more sophisticated and slower process would analyze all the information provided from our senses and correcting the spatial reference frame known from the first process.

Different senses provide either absolute or relative information which could be integrated into the frames of reference. **Vision** is known to provide us with excellent measurements about our location and movements in space. Distance to objects, their movement direction and relative speed can be easily extracted in hardly more than the blink of an eye. Optic flow can give us the sense of self-motion, if the visual field of view is large enough. It can be used to navigate through an environment (Beall and Loomis, 1997; Riecke, 1998). Landmarks, in contrast to optic flow, are more reliable and do not lead to accumulation errors (Bülthoff et al., 2000). Spatialized **sound** is worse than vision; front-back confusions and a poor spatial resolution produce an unreliable spatial location (Begault, 1994). However, there is an auditory reference frame, which could help to focus our attention towards targets and localize them roughly in space. Disturbances of this rather inaccurate reference frame, achieved by unnaturally changing sound sources, may nonetheless confuse our belief in one coherent or unique reference frame for all the senses. **Proprioception** provides information about our body position. Active and passive changes to body limbs are "known" to the system and contribute to motion perception (Mergner et al., 1991; Hlavacka, Mergner, and Bolha, 1996). Also, the force on different joints introduces knowledge about our position relative to external forces (e.g. gravity). Finally, the **vestibular system** is known to measure changes in velocity by means of linear and angular accelerations. We maintain a strong sense of gravitational direction, which is known with respect to our body axes (Mergner and Rosemeier, 1998). If the sensation of gravitational force is unpredictable or unstable, we have trouble in maintaining straight gait or posture.

So far it is unknown if and how these different frames of reference are integrated to provide one unique and stable reference frame. One could ask how important these sources of information are to our perceived location in space. One approach is to determine each sensory modality's ability to provide a reliable perception of our spatial location as some of the above mentioned studies did. The other extreme would be to try to control the input provided to all senses at once and ask, by changing small parts of the system, how much those changes influenced the perceived location in space. Of course, there are lots of ways that could be attempted. Here, we have chosen to determine the information provided by the vestibular system to our location in space. Naturally, we could not stop the other senses from providing information. However, we made that information as useless (uncorrelated to the task) as possible.

Coming back to the original question of one's perceived location in space, our approach is to divide the question in different steps with increasing complexity:

- **What do people perceive when they are passively moved and have vestibular input only?**

  How does it feel to be moved in space? How do we maintain a stable representation of the environment if we do not perceive these changes visually? Is the vestibular input sufficient to maintain a stable frame of reference? Moreover, does the vestibular input provide enough information about changes in our own position in space to force an automatic update of the spatial reference frame?

- **Can people distinguish between accelerations, velocities, and distances?**

  Even though the vestibular system is known to measure acceleration, one could ask if humans perceive maximum velocity and changes in position. These values could be derived from the acceleration signal by mathematical integration over time. When people are asked to judge their movement in space, can they separate those values? Is there a distinction between perceived distance, velocity and acceleration?

- **Can these values be estimated on a relative scale?**

  If there is a conscious correlate of acceleration, velocity and distance, could one make a judgement regarding the strength of the value? Is there a relative scale for those values and could this be verbalized? Can humans judge these vestibular signals as acceleration itself and integrate them to reliably derive velocity and distance estimates?

The questions above were investigated with the following psychophysical experiment. To focus on the perceived vestibular input, we have to disrupt all other spatial cues, while carefully controlling the vestibular input.

3.1 Experimental design

Twelve healthy subjects (six men and six women), randomly chosen from the MPI subject database, were asked to perform a series of experiments. The age of the subjects was between 19 and 29. Subjects gave their informed consent and were paid for their participation. The experiment was approved beforehand by the local ethics committee.

As the focus of this experiment was on the information provided from the vestibular system, all other external spatialized cues were eliminated. The subjects performed the task blindfolded. Spatial auditory cues were effectively diminished by the noise canceling headphone system described in section 2.2.7. In addition, non-spatial broad band noise was played through the headphones. The subjects were passively moved and therefore had no efference copy of any motor commands. The position of all body parts was known to the subject, but was not changed during movements. We eliminated movement specific vibration cues from the motion platform by adding independed white noise in all six DOF to the movements provided. The spatial location during the experiment was therefore primarily determined by the vestibular stimulation described in the next paragraph.

The vestibular input was designed to allow independent variation of the distance traveled and the maximum acceleration reached without correlation to the duration of the stimulation. The stimulation time was held below four seconds to avoid adaptation to one acceleration level. The motions were determined by Gaussian-shaped velocity profiles. The widths of the Gaussian curve were chosen to create a full two-factorial design of five distances and six maximum accelerations. Due to the natural restriction of the combination of distance, velocity and acceleration, the maximum velocity reached varied as a result of the other two independent factors. However, for certain factor combinations, similar maximum velocities were reached (see table 3.1 and 3.2). Figure 3.1 shows all combinations of the five distances and the six accelerations with their resulting motion profiles. The profiles were performed for linear movements (X:forwards-backwards and Y:left-right) and turns around the subject's vertical axis (H:heading).

	distances[m]				
accelerations[m/s^2]	0.05	0.10	0.15	0.20	0.25
0.1	0.057	0.081	0.100	0.115	0.129
0.2	0.081	0.115	0.141	0.162	0.182
0.4	0.115	0.162	0.199	0.229	0.257
0.6	0.141	0.200	0.244	0.281	0.314
0.8	0.163	0.230	0.281	0.325	0.363
1.6	0.231	0.326	0.398	0.460	0.514

Table 3.1: Reached maximum velocities [m/s] for all combinations of the five distances and the six accelerations for the linear movements.

The vestibular stimulation was performed by the motion platform of the Motion-Lab with an update frequency of 400 Hz based on a profile defined at a rate of 30 Hz which results in smooth movements. The platform control program used a standard second order low-pass filter to make a smooth movement following the given positions over time. To avoid a noticeable change in platform vibration that could be correlated to a certain velocity, random noise jitter was superimposed on the motion profile. The jitter was uncorrelated in all six degrees of freedom. The linear range was ±2mm and the angular component

	angles[°]				
accelerations[$°/s^2$]	5	10	15	20	25
10	5.7	8.1	10.0	11.5	12.9
20	8.1	11.5	14.1	16.2	18.2
40	11.5	16.2	19.9	22.9	25.7
60	14.1	20.0	24.4	28.1	31.4
80	16.3	23.0	28.1	32.5	36.3
160	23.1	32.6	39.8	46.0	51.4

Table 3.2: Maximal turn velocities [$°/s$] for the all combinations of the five turn angles and the six angular accelerations for the rotatory movements.

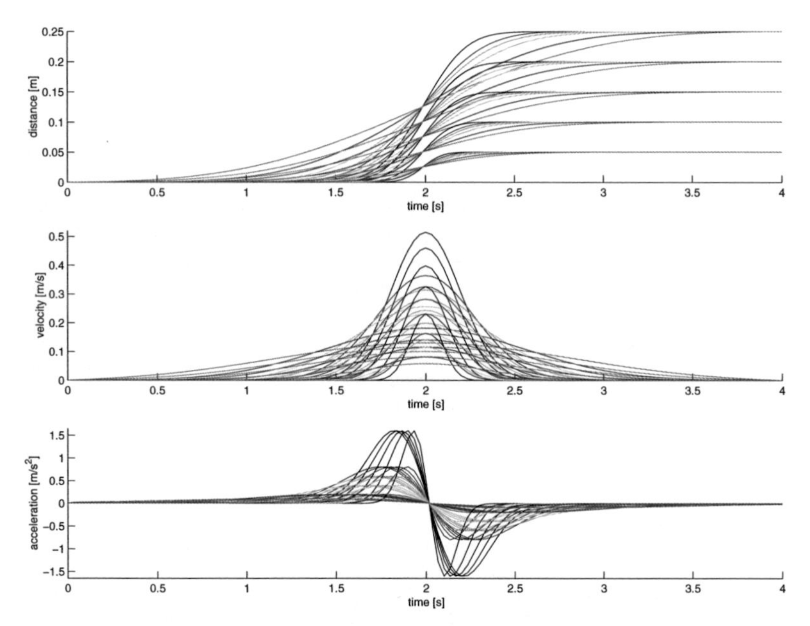

Figure 3.1: The upper graph displays the change in position (distance traveled). The middle graph shows the corresponding velocity profiles. Finally, acceleration is plotted in the lower graph. The translation profiles were created from the Gaussian-shaped velocity profiles. The acceleration describes a function close to a sinusoidal profile. The plot shows the factorial combination of five distances (5, 10, 15, 20, 25 cm) and six accelerations (0.1, 0.2, 0.4, 0.6, 0.8, 1.6 m/s^2). Profiles with the same maximum peak acceleration are plotted in the same color.

was restricted to $\pm 0.16°$. The same amount of jitter was used for all the experimental conditions.

The subjects' task was to judge the presented movements on a scale between 1 and 100%. They were instructed to look for the maximum stimulus (=100%) and rate the other stimuli in comparison. On this scale of 1 to 100%, the subjects had to verbally judge in three blocks the distance traveled, maximum velocity reached and maximum acceleration reached. In addition, the subjects judged the direction of motion. Each block consisted of all factorial combinations (5*6=30) in both directions with four repetitions in random order (30*2*4=240). Due to the limited motion range of the motion platform, the random order was rejected beforehand when the stimulus sequence would exceed more than ±0.45m or ±45°.

female	male	session order	feature order		
			X	Y	H
SKW	WPI	H X Y	a v d	v d a	d a v
AXF	CFB	Y H X	a d v	v a d	d v a
RTH	PDO	X H Y	d a v	a v d	v d a
GME	EKO	H Y X	d v a	a d v	v a d
HYD	TPM	Y X H	v d a	d a v	a v d
FMC	KIW	X Y H	v a d	d v a	a d v

Table 3.3: Experimental order of conditions. In column session order: X = forward-backward, Y = left-right, H = heading left-heading right. In column feature order: d = distance, v = velocity, a = acceleration.

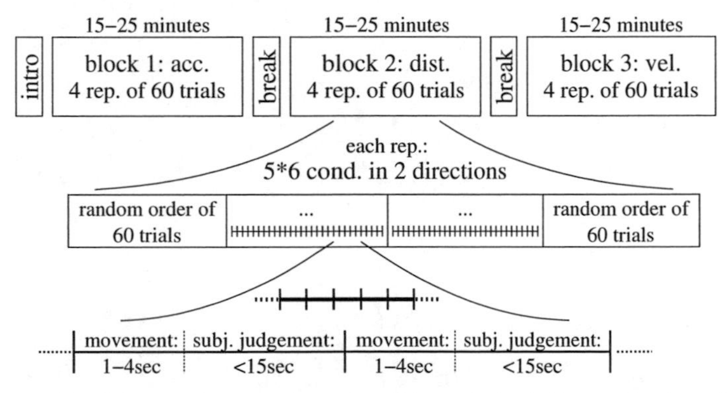

Figure 3.2: This example shows the order of blocks for the X (forward-backward) condition for one subject (AXF). After a short introduction to the purpose of the experiment and the task, this subject was asked to judge acceleration in the first block. These 240 trials (60 trials in four repetitions) were followed by a break and then the next two blocks with a break between them. The 60 (=2*5*6) trials of each repetition were randomized. Each individual trial had a maximum duration of four seconds followed by a brief period during which the subject gave the verbal judgement to the experimenter.

During one block, the subjects had to judge only one value (distance, velocity or acceleration) per trial, plus the individual direction they were moved in. The movement itself was the same for the different judgement tasks, but the subject had to concentrate on a different stimulus feature. The order of tasks (distance, velocity, or acceleration judgement) was varied across subjects. Between these blocks, the subjects usually took a short break. Different movement directions (forward-backward, left-right, heading turns) were performed in three separate sessions usually on different days. One session lasted $1\frac{1}{2}$ hours on average for the subjects. In total, each subject had to judge 5*6*2*4*3*3=2160 (distances * accelerations * directions * repetitions * features * degrees of freedom) movements[1]. Subjects were randomly assigned to the different conditional permutations (see tab. 3.3). The subjects names are printed in an anonymous form of random three letter combinations (e.g. SKW). To clarify the experimental design, figure 3.2 shows an example of one session.

3.2 Results

Naturally, the results of this experiment could be plotted in a number of ways showing different human abilities. However, this section tries to focus primarily on the questions given above. This is done in two parts. On the one hand, it is useful to look at the individual performance of different subjects and look for similarities in their performance pattern. On the other hand, it makes sense to pool data across subjects to gain results which hold for the whole population. Nevertheless, the pooled results should be similar to most individuals in order to avoid artifacts.

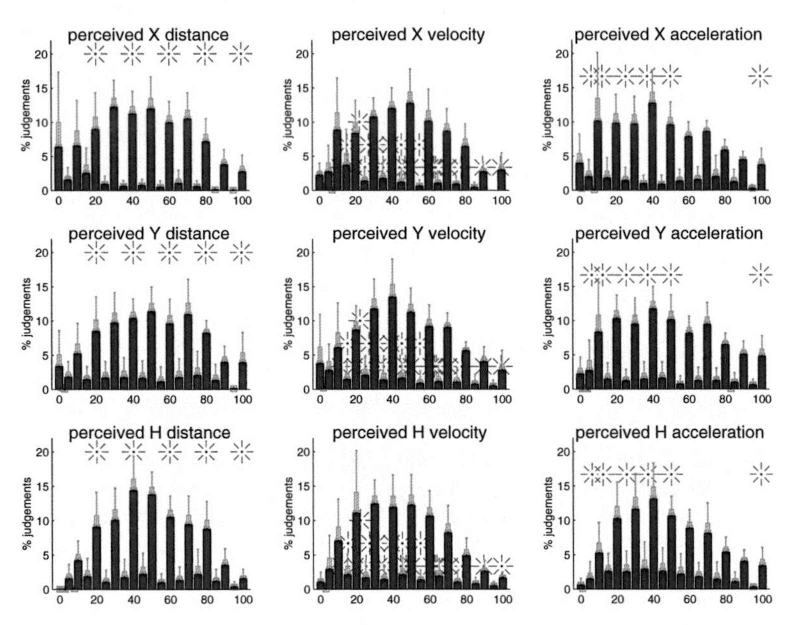

Figure 3.3: The histograms of subjects' responses is depicted with red bars for the respective condition. The green bars refer to the standard error of the mean, with "whiskers" depict one standard deviation. The blue "stars" show the distribution of the theoretical perfect answer.

[1]In total, all sessions took more than 60 hours of experimental time, in which 12*2160=26140 data points were collected.

In general, subjects had problems adopting the method of using a scale of 1 to 100. Some subjects did not use the whole scale during the experiment; they always stayed below the maximum range they could and should have used, indicating that they did not classify one of the stimuli as largest. To eliminate the beginning phase of adaptation to the scale, we excluded the first of the four repetitions from the further analysis. While this adaptation is not a general problem for the individual data, it might introduce variability into the pooled data. Even when subjects used the whole scale, they tended to use the middle range more frequently. This tendency is very often observed, especially for tasks similar to ours (Poulton, 1981a; Kowal, 1993), but also for tasks were a non-verbal response was measured (Berthoz et al., 1995; Ivanenko et al., 1997a). Figure 3.3 shows the histogram of all subjects' responses for the 1 to 100 scale. Besides a general tendency towards a Gaussian distribution, all factors of ten got used more often than anything in between. Some subjects made use of more than 10 values, but anything below a differentiation of 5% was very rare. The figure also depicts the percentage of theoretical right answers across the stimulus range with blue "stars". Interestingly, the distribution of right answers had no effect on the distribution of the subjects' responses, showing that the difference between a linear distribution (distance estimates) and a exponential distribution (acceleration judgements) was not affecting the use of the scale.

Overall, the judgements of movement direction showed no asymmetry for the different directions, which enabled us to pool the data across both directions. The movement direction judgements are interesting only for very slow or short movements, since only in those trials did subjects make errors. This indicates that the vestibular input for most of the trials was sufficient to let the subjects know in which direction they were moved.

One concern during the design of the experimental trials was that subjects could answer by pure time estimates of the moved time instead of the asked values of distance, maximum velocity, and maximum acceleration. This assumption was proven to be wrong by calculating the correlation between the stimulus duration with an acceleration above perceptual threshold and the respective answers to the questions. The correlation between the theoretical "right" answer and the movement duration was not negligible (distance: $t(58)=5.31$, $p<0.00001$***, $r^2=0.33$; velocity: $t(58)=3.06$, $p=0.0017$**, $r^2=0.14$; acceleration: $t(58)=7.17$, $p<0.00001$***, $r^2=0.47$). As we had hoped for, the correlation of the responses of all subjects with the respective stimulus duration were less strong. Nonetheless, the subjects judgements showed high correlations to stimulus duration, but the r^2 values stayed on a lower level (distance: 0.06-0.13; velocity: 0.06-0.12; acceleration: 0.15-0.28). Moreover, two of the subjects were later asked to explicitly judge the duration of their movement and reached a high correlation with high r^2 values ($t(178)=14.2$, $p<0.00001$***, $r^2=0.53$ and $t(178)=12.2$, $p<0.00001$***, $r^2=0.46$).

3.2.1 Individual data

Individual data look, in general, more noisy than the pooled and preprocessed data. Nonetheless, this section starts with a look into the raw data and explains certain things one can already see in those plots. Subjects judged the distance, velocity, and acceleration quite consistently, but with systematic errors. Figure 3.4 plots the data of the left-right translation block for a typical subject (EKO). The data is plotted multiple times, but against different axes and grouped for different conditions. Each individual plot shows the data twice: on the left hand side it is grouped by identical maximum acceleration and on the right hand side by identical distance. This allows a differentiation between the effects of the various factors of the experimental design. For comparison, the theoretical perfect answer is plotted in Fig. B.1 in appendix B.

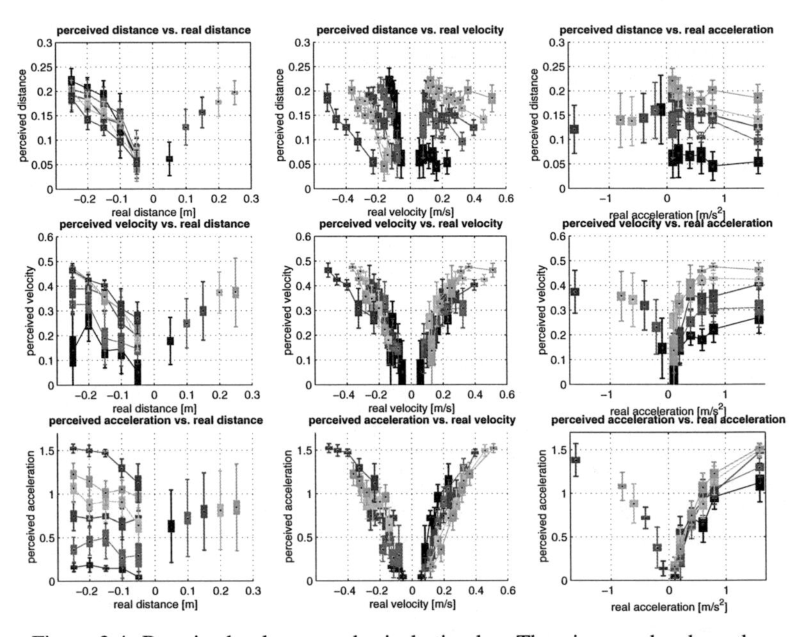

Figure 3.4: Perceived values vs. physical stimulus: The nine graphs show the data of a typical subject in the Y (left-right translation) condition. The rows contain the data for distance, velocity, and acceleration judgement, respectively. Arranged in columns, the data is plotted against the physical distance, velocity, and acceleration of the stimulus. Therefore, the diagonal displays the subject's answers correlated to the physical stimuli. Each plot displays the data twice: the left-hand side shows the data for identical peak acceleration and the right-hand side groups the same data for identical distance.

Several points can be observed in the plots. The sub-graphs on the diagonal show a strong correlation between the physical stimulus and the subjects' judgements: distance: $t(178)=18.2$, $p<0.00001$***, $r=0.81$; velocity: $t(178)=15.2$, $p<0.00001$***, $r=0.75$; acceleration: $t(178)=22.3$, $p<0.00001$***, $r=0.86$. This indicates that subjects accurately judged the appropriate stimulus feature. Further, the distance estimates, for example, show the same tendency towards the mean which was already observed in the histogram plot in Fig 3.3. On the other hand, plotting the perceived distance against the physical acceleration shows a good independence of the judgements from this factor. In addition, the distance estimates are weakly correlated with the real velocity but negatively correlated with the acceleration, as shown in the upper row, middle and right graphs (vel: $t(178)=2.83$, $p=0.0026$**, $r=0.21$) (acc: $t(178)=3.17$, $p=0.00089$**, $r=-0.23$). However, one general observation is the similarity of the second and third row of the plots. The acceleration judgements show weak correlations to distance ($t(178)=2.23$, $p=0.013$*, $r=0.17$), but strong correlations to the physical velocity ($t(178)=22.6$, $p<0.00001$***, $r=0.86$), which may indicate problems to judge maximum accelerations independent from peak velocity. The similarity could be caused by a misunderstanding of acceleration in contrast to velocity. The alternative explanation would be that humans can not judge the peak acceleration independently from the maximum velocity. A comparison of the results of physics students with others students suggests that the first alternative is correct.

3.2.2 Linear model fit

Another approach to dissociate the impact of the different physical values on the actual judgements would be to calculate a linear model that fits the judgements based on the real physical values. A model which assumes a linear combination of distance, velocity, and acceleration was fitted to the individual data with minimal error.

$$J_i^C = D_i * Distance + V_i * Velocity + A_i * Acceleration + Error_i$$

For each experimental condition $C \in \{distance, velocity, acceleration\}$, the coefficients (D_i, V_i, A_i) describe the subjects' (i) response J_i across all 30 factorial combinations. The error was minimized for each subject independently for the combination of the three coefficients D, V, A. The result of this fit to the same data as shown in Fig 3.4 is now plotted against the two factor axes in Fig 3.5. The first row of graphs shows the subjects' response for the 30 factor combinations coded in color. The three plots contain the data for distance, velocity and acceleration judgements, respectively. In this kind of plot the perfect distance judgement would result in a horizontal grading which changes only by the distance factor. In contrast, the acceleration estimate should result in a vertical grading increasing with the acceleration axis from left to right. Finally, the velocity plot showing the combination of the two other factors would not be a linear grading, but curved: One could reach a certain maximum velocity with slow acceleration and large distance or with a faster acceleration but shorter distance. However, the same velocity is reached by different combinations which do not fall on a straight line in this kind of plot.

Figure 3.5 depicts in addition to the subjects' responses, the model fit, and the difference (error) between them. The error again shows the tendency towards the mean response by adding a non-linear compression to the data which was not resolved in the linear models fit. In comparison to the theoretical perfect answer described above, it has to be noted that the model fitting subjects' responses confirm the analysis from above: The acceleration judgement is very close to the velocity judgement. Nonetheless, a tendency can be observed emphasizing the acceleration coefficient in comparison to the distance coefficient for the acceleration judgements. However, the pooled data is of greater interest, especially for the model fits.

3.2.3 Pooled data

Pooling of data is useful to indicate and emphasize general features of the data which were found already in the individual data. The pooled data will generalize across individuals and allow us to speculate about general findings which should hold true for the rest of the human population. It also enables us to see the individual variability of the results. Different individual variances were eliminated leaving the inter-individual differences for the analysis.

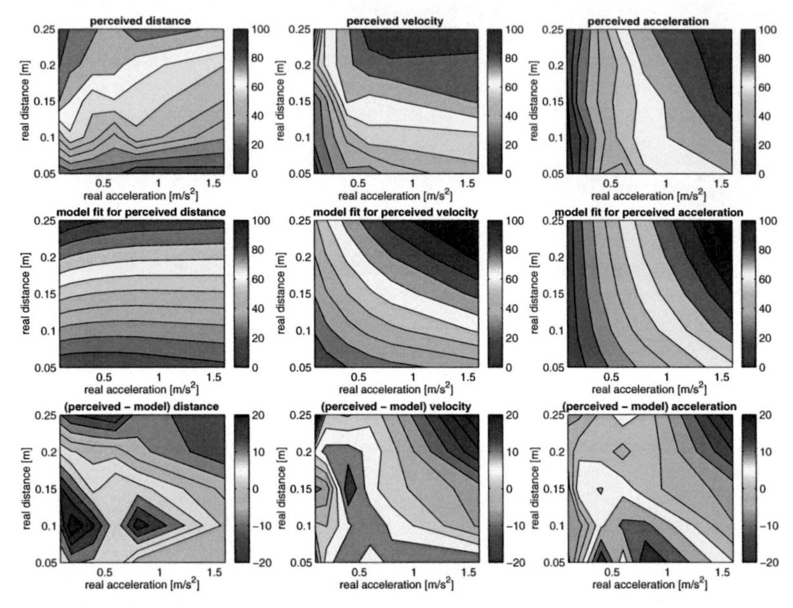

Figure 3.5: Model fit to individual data: The nine graphs show the same data as Fig. 3.4. This time, each plot in the first row displays one block of the experiment. The color encodes the mean of the subjects' judgements across all factorial combinations. The second row is the result of the fitted model to the individual data of the first row. The last row depicts the difference between the model and the actual data. Note the different color scales for the different rows allowing to emphasize the small error.

The histogram plot in Figure 3.3 has already shown the similarity of the subjects' usage of the estimation scale; it should be emphasized that this underlines the strong agreement of the subjects' verbalisation of their perception. If one of the subjects did not understand the scale, did not use it or had unusual perceptions of the movements due to a hidden vestibular deficit, this would turn up in this plot. However, none of the subjects had to be excluded from the analysis.

In appendix B, more graphs show the pooled data in a similar form as Fig. 3.4. In Fig. 3.6 subgraphs from Fig. B.2 to Fig. B.4 are extracted to point out some more details. The graphs plotting perceived distance against acceleration show small differences across the used DOF. For the linear DOF (X and Y) clear saturation curves show a threshold reaching a plateau around 0.3 $[m/s^2]$. This threshold might be higher than other values reported in the literature due to the added white noise. The graph for turns (H) show nearly no saturation due to the different perceptual accuracy of the canal system in comparison to the otoliths. Comparing the plots of perceived distance against distance pooled for acceleration, one could see that all of them show a grading. The graphs arrange themselves starting with the lowest acceleration toward the highest getting closer to a "perfect" judgement, meaning getting close to a linear relation of $1 : 1$. Again the graphs of the linear DOF's show a higher similarity. An other interesting group of plots depicts the perceived maximum acceleration against real maximum velocity. Here the lateral movements (Y) differ

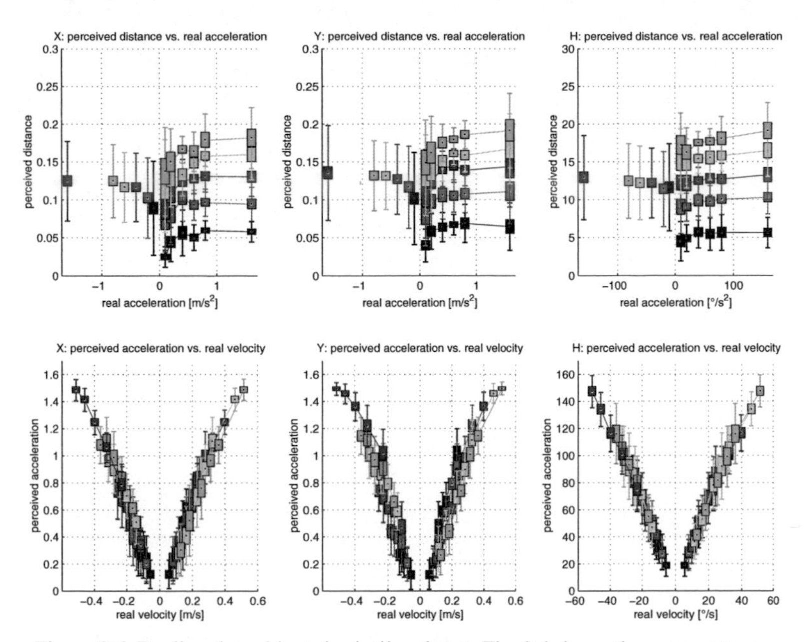

Figure 3.6: Pooling the subjects in similar plots to Fig. 3.4 shows the same pattern as described for the individual subject in the previous section. Here, the graphs show a subset of all the plots from the appendix B.

from the others: One can see the tendency to spread out more towards the perfect performance, which should show an independent horizontal line here (compare with Fig. B.1). However all three plots show that the acceleration judgement is highly correlated with the velocity value.

DOF	task	D	V	A	Error
X	D	0.632	0.409	-0.066	0.620
	V	-0.019	1.174	-0.168	0.440
	A	-0.115	1.112	-0.006	0.396
Y	D	0.677	0.295	-0.009	1.115
	V	-0.011	1.069	-0.076	0.574
	A	-0.073	0.896	0.160	0.669
H	D	0.847	0.067	0.057	0.658
	V	0.161	0.887	-0.065	0.452
	A	0.045	0.947	-0.010	0.328

Table 3.4: Coefficients of the model across all subjects. The coefficients were calculated on the average responses from the individual subject.

Pooling the coefficients of the model described earlier results in the plots of Fig. 3.7. The deviation from the mean coefficients shows to some degree the individual differences in the strategies used. For some subjects, the coefficients for judged velocity and acceleration

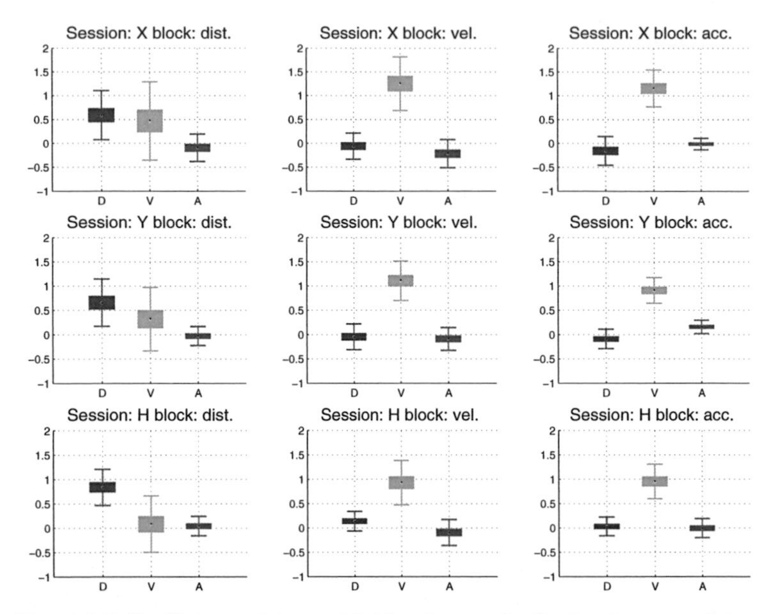

Figure 3.7: Coefficients of the model: The nine graphs display the mean of the coefficients from the model (see Fig. 3.5) for all nine experimental conditions. The rows show the three sessions: X (forward – backward translation), Y (left – right translation), and H (turn around the body's vertical axis). The columns show the different experimental blocks of each session (distance, velocity and acceleration judgement). The bars refer to the standard error of the mean, and the "whiskers" depict one standard deviation.

differed and for some others there was hardly any difference. As mentioned earlier, this fact might be caused by a misunderstanding of the term "acceleration" or, as it was described to the subject, the "force" which was acting upon them[2]. Regardless of the similarity observed between the velocity and acceleration judgements, there is a slight difference between those judgements even for the pooled data: The ratio between the distance and acceleration coefficients changes between distance, velocity and acceleration judgements. The portion explained by the acceleration increases, whereas the portion explained by the distance decreases from distance to acceleration. This ratio is also depicted as the tangent to the color grading in Fig. B.5 to Fig. B.7[3]. The slant of the tangents on the graphs' diagonals changes in correlation to the ratio between distance and acceleration coefficient, since the coefficient for velocity does not change the tangent on the diagonal. Additionally, it is important to mention that the pattern of coefficients for the different degrees of freedom tested show a high similarity, suggesting a common base for the judgements of translational and angular movements. Table 3.4 shows the exact numbers for the model fit to the averaged individual data. The individual data is pooled ignoring the directional information. The averages therefore contain no individual variability. The model fits the average value with minimal error.

[2]Of course the force is proportional to the acceleration used: $F = m * a$.

[3]All three of them being in appendix B

3.3 Discussion

We started the experiment with three questions in mind which will be discussed in the following. Since the experiment was carefully designed not to provide any other reliable cue than vestibular stimulation, we asked subjects to report their subjective judgement of distance, maximum velocity, and maximum acceleration.

- **What do people perceive when they are passively moved and have vestibular input only?**

  Subjects reported verbally on a scale of 1 to 100 indicating their subjective scale of distance, maximum velocity, and reached peak acceleration. They perceived the changes based on vestibular and somatosensoric input and had to maintain a stable representation of the environment in order to judge the distance traveled relative to the environment. The judgement was effortless, since most of the subjects could report spontaneously directly after the trial. The verbalization seemed to be unnatural in the beginning, but subjects very quickly got used to the task and answered sometimes during the trials (especially when prompted for maximum values).

- **Can people distinguish between accelerations, velocities, and distances?**

  The subjects reported different judgements for the different tasks using the same profiles. This indicates that they distinguished between the values in question. In order to come up with a distance correlated judgement, the subjects had to perform a double integration of the vestibularly perceived acceleration signal. The actual acceleration judgement was very close to the velocity judgement. This can be interpreted as an inability to regenerate the acceleration from the perceived velocity signal from the vestibular system. An alternative explanation is that the concept of acceleration is somehow not clear to most of the subjects, since only some of them showed clear distinctions between the two judgements.

- **Can these values be estimated on a relative scale?**

  The subjects performing relative judgements used most of the available scale. A relative judgement of a movement to a globally perceived maximum was possible. The derived values were stable over the course of the experiment shown by a small variance for multiple repetitions of the same factor combination. This indicates that a memory for motion profiles exists which can be used to compare multiple movements for the parameters in question. The conclusions of experiments where subjects reproduced passively perceived profiles indicated that solely the velocity profile was stored (Berthoz et al., 1995) can not be confirmed. At least it does not seem probable that two velocity profiles are effortlessly compared with respect of their integral. It seems more natural that the actual distance of the trials was perceived and memorized for comparison.

As an extension to the study by Mergner et al. (1996), which found that perceived angular displacement is the time integral of perceived angular velocity, we can add that it seems likely that this is also true for linear movements. The comparison between linear and angular judgements does not indicate a disjunct processing of the cognitive estimates of the separated sensor organs. It appears as if both, otoliths and canals signals, are handled with the same brain structure or at least same mechanism in order to judge the parameters we asked for. The similarity of the judgements does strongly suggest a common representation in sense of spatial updating.

Since Jürgens, Boß, and Becker (1999) proposed a common structure for the active reproduction and the base for the verbal judgement of turned angle (= total angular displacement), one can speculate about extending this notion to linear movements. Jürgens et al. found no difference for their comparison of active and passive condition. In the active condition, subjects were asked to perform a turn of given total displacement, and in the passive condition they verbally judged the total displacement of turn movements. Compared to our results, one would expect that subjects should have been able to turn 50% of their perceived maximum as good as they actually judge the same movement. Extending this to linear movements, one might expect subjects to be similarly precise in actively producing distances as they were in performing verbal judgements in our experiment.

In sum, the subjects estimated peak velocity and traveled distance, which they obtained from vestibular perceived acceleration. It remains unclear why the acceleration judgement looks like a velocity judgement. One possible explanation is that since velocity is encoded at least for angular movements at a very early stage in the sensory organ, it may replace or mask the acceleration signal from later cognitive access. In total, the vestibular system is providing a good spatial reference frame which does enable spatial updating. This leaves the question if vestibular simulations for bigger scale movements is possible at all on this kind of motion platform.

Chapter 4

Experiment 2: Holding balance, and coding vestibular and visual heading changes

This experiment focuses on two rather automatic functions performed by humans in every day life: Maintaining balance and the perception of turns. Specifically, humans are used to maintaining balance while walking or standing, a skill that develops early in childhood. This automatic behavior needs little conscious control once we have learned and practiced it for years. Similarly, the perception of ego-turns seems to need little conscious effort. Nonetheless, binding a person's mental activity can cause disorientation and disturb the perception of turns, as Yardley and Higgins (1998) pointed out. The perception of turns is crucial for navigation and self-localization. The underlying processes automatically transform ego-turns into corresponding changes of our spatial reference frame.

The latter is not as obvious, but very simple to demonstrate: ask a standing person to point towards a landmark not too close by - let the person turn by 90° with eyes closed - ask the person to point quickly to the remembered position **ignoring** the turn just performed. Most people will point to some oblique angle but not in a direction which is 90° off from the actually remembered direction. Instead of performing the exact same motor command while pointing, some updating process occurs which is hard to suppress, changing the remembered direction. Further, if one asks people to point towards the actual landmark after performing the turn, their accuracy is remarkably high. We call this updating process *obligatory spatial updating*.

Several cues are known to be sufficient for the perception of turns and other basic navigation tasks. Yet in most VR applications, turns are misperceived (see Bakker, Werkhoven, and Passenier, 1999) and lead to disorientation. This might be due to the fact that only some of the cues available in the real world are simulated. In addition, the cues which are simulated and presented are imperfect or restricted, like the limited field of view inside an HMD. A small mismatch or even conflict between cues might deteriorate the percept and destroy the ease of the automatic spatial updating. However, in some experimental conditions naïve subjects seem not to notice any mismatch (Ivanenko, Viaud-Delmon, Siegler, Israël, and Berthoz, 1998).

The same senses which contribute to the perception of turns are also involved in maintaining one's balance: The vestibular system provides a sense of gravity direction and angular velocity, but no absolute reference frame; proprioception, mainly from the feet, provides pressure information and information about the body joints; vision in general provides a very stable and absolute reference frame. Presenting solely optic flow reduces vision so that it also provides no absolute reference frame.

Regarding linear and angular movements separately, others have conducted experiments similar to our experiment (which has not been described yet). Israël, Sievering, and Koenig (1995) performed experiments where subjects had to estimate their self-rotation actively on a joystick-controlled mobile robot. In Berthoz et al. (1995), subjects were asked to reproduce passively learned linear displacements (2 to 10m) on the same mobile robot (without vision). Active and passive turns were compared by Jürgens et al. (1999): Blindfolded subjects had to judge and execute active as well as passive turns, while either freely standing or standing on a turn table.

In contrast to the above mentioned studies, which were done without visual cues, in our experiment we used two cues that provide no absolute spatial reference (optic flow and vestibular cues). We performed a series of experiments concentrating on the following questions:

- **Can humans maintain balance without vision?**

  Standing in the dark is possible for most humans using proprioception in the feet and the vestibular system. Transferring the balance problem to *seated* subjects reduces the influence of proprioception. Can humans nonetheless control balance and maintain an upright position?

- **Is it possible to ignore additional rotations which are not related to the task?**

  Is it possible to add rotations in a different degree of freedom without disturbing performance? If visual feedback provides the same information as the additional vestibular rotations, is that visual information interfering with the primary balance task?

- **Which heading information is coded during path following?**

  Can humans learn and remember the additional rotations? Is the vestibular and/or the visual information used to reproduce the correct path heading changes?

- **Is the information differently coded when being moved passively?**

  The active control of balance might have further impact on the coded information of the turn amplitudes. Can humans extract the same information about the heading changes when being moved along the virtual path without active control of the movement? How does the difference between active/passive condition change the memorized turns?

4.1 Methods and experimental design

We used the Motion-Lab VR setup including the motion platform and the HMD for presenting vestibular and visual stimuli, respectively. (See chapter 2 for detailed descriptions of the equipment.) Vestibular cues consisted of head centered heading and roll rotations. Visual self motion was presented as optic flow. During each trial, the subject had a constant visual forward velocity of 1 m/s. Spatial auditory cues were excluded by noise canceling headphones. Subjects interacted with the simulation via a joystick.

Eight healthy subjects (four male and four female with ages between 16 and 33) were asked to perform the experiment in the active observer group. Seven other subjects (four male and three female with ages between 18 and 45) participated in the passive condition. Subjects gave their informed consent and were paid for their participation. The experiment was approved beforehand by the local ethics committee.

On average, the active condition experiment lasted about 1.5 hours; the passive condition had a duration of only 45 minutes. In the active condition, subjects learned to maintain an upright position before the actual test phase. This training section was not present for the passive condition, since subjects did not have to control balance in this condition. In the test section of the experiments, subjects had to memorize turns and reproduce them actively with the joystick. The overall course of the experiment is described in Fig. 4.1.

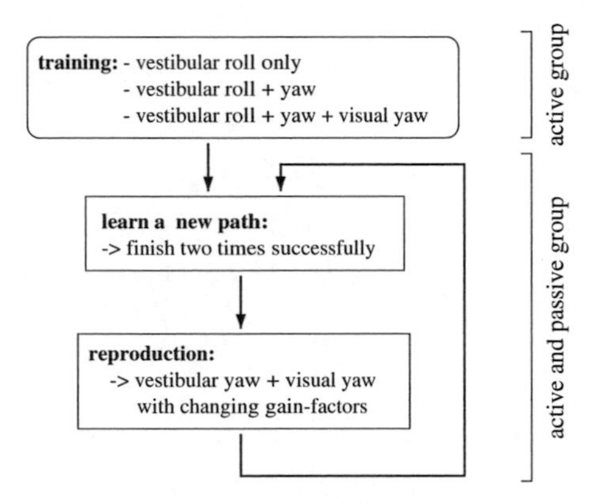

Figure 4.1: **Training (active group only):** the subject learns how to handle the joystick and control the roll stabilisation of the platform. In three short blocks the training complexity increases. **Test-phase (both groups):** the subject learns a new path until two successful completions and has to reproduce the learned yaw rotations with changing gain-factors between visual and vestibular turns.

In the training section subjects of the active group were asked to maintain an upright position by continuously adjusting to changes in the platform's tilt. In fact, the subject was following a path which was randomly generated and included heading changes between 8.5 and 17 degrees. Deviations from this path were presented as vestibular roll rotation. Therefore, to maintain balance, the subject had to follow the path, controlling his/her heading with the joystick. The training section was split into three parts with increasing complexity. In the first two parts no visual optic flow was presented. Subjects were blindfolded and trained on the balance task. The first part presented roll rotations only, whereas the second part added the actual heading changes made. The third training part presented visual flow for heading changes matching the yaw rotations of the platform. It is important to note that the path was solely defined by vestibular roll and was never visible. Moreover, during the training phase the relationship between vestibular and visual turns was always fixed, i.e. the gain factor was 1.0. The complete training lasted about 30 to 45 minutes.

For all subjects, the test section was divided into two alternating phases: The learning and the reproduction phase. The ability of the subjects in the active group to maintain balance and thereby follow a predefined path was used here. The subjects of the passive group followed the same path by computer control. After successfully following a vestibularly defined path twice, subjects were asked to reproduce it (i.e., the sequence and amplitude of the three turns) from memory. The joystick was used to control the angular velocity of the turn. During the reproduction phase this transformation was varied by a gain factor of $1/\sqrt{2}$, 1 or $\sqrt{2}$ for each modality separately. Therefore, the ratio of visual and vestibular turns was between 0.5 and 2.0. Overall, subjects were asked to learn nine randomized paths including turns of 8.5, 12 and 17° and reproduce the memorized turns with three different gain factor combinations each. See table 4.1 for the summary of heading change and gain factor ratios. Therefore, the data of each subject contains three repetitions for each angle over all nine factor combinations. This part of the experiment lasted about 45 to 60 minutes.

The active versus passive manipulation was introduced to determine whether subjects of the two groups would code the learned path differently. The subjects of the passive group learned the turns without experiencing any particular gain factor for the joystick control. Therefore, they could not simply reproduce the motor pattern necessary for the heading changes.

	gain factor		
heading change	$\frac{1}{\sqrt{2}}$	1.0	$\sqrt{2}$
8.5	6.0	8.5	12.0
12.0	8.5	12.0	17.0
17.0	12.0	17.0	24.0

	vestibular gain		
visual gain	$\frac{1}{\sqrt{2}}$	1.0	$\sqrt{2}$
$\frac{1}{\sqrt{2}}$	1	$\frac{1}{\sqrt{2}}$	$\frac{1}{2}$
1.0	$\sqrt{2}$	1	$\frac{1}{\sqrt{2}}$
$\sqrt{2}$	2	$\sqrt{2}$	1

Table 4.1: This table shows the gain factors used for changing the relationship between heading of the virtual path and the vestibular or visual turns executed. The gain factors themselves and the used heading changes have a relation of factor $\sqrt{2}$ between them. Combining two gain factors results in a ratio of 0.5 to 2.0 between executed visual and vestibular turn.

4.2 Results

The subjects' performance was analyzed separately for the two main parts: training and test phase. During the experiment data was continuously recorded at approximately 100Hz allowing the analysis of single trials off-line. The analysis of a single trial looked for reversals in the heading direction, so that the turn amplitudes of the performed paths could be extracted. When two curves in the same direction follow each other directly, this method will extract the total turn amplitude. During the training phase, up to two consecutive turns could have the same direction. However, in the test phase the curves were always changing direction.

During the experiment, no subject had to stop due to simulator sickness, despite the large mismatch between vestibular and visual turn. Most of the subjects actually did not note any mismatch even when specifically asked after the experiment. One subject reported spontaneously a strong difference in vestibular and visual turn and reported further not to have payed much attention to the visual stimulus from that point on. As later analysis showed, the visual gain factor nonetheless influenced this subject's response. Since the subjects always learned a new sequence of turns in a no conflict condition, it is not expected that subjects would adapt to any mismatch of the gain factors. This was confirmed by the average of the response not changing in the course of the experiment.

4.2.1 Training phase

In general, subjects of the active group quickly learned to control their upright position with the joystick. Slowly increasing the speed and amplitude of the disturbances brought subjects to a comparable performance level. The level of performance can be analyzed by means of deviation from the upright position subjects were asked to maintain. All samples (at 100 Hz) were taken for each individual training trial and the frequency of deviation from vertical position was calculated in bins of 1° width. The results can be plotted in response frequency over the deviation from the middle position and fitted by Gaussian shaped curves. Taking all trials per subject result in a pooled accuracy curve. Figure 4.2 shows the average deviation from the predefined path resulting in an average roll angle of -0.17° with a standard deviation of 0.66°. This is not significantly different from zero.

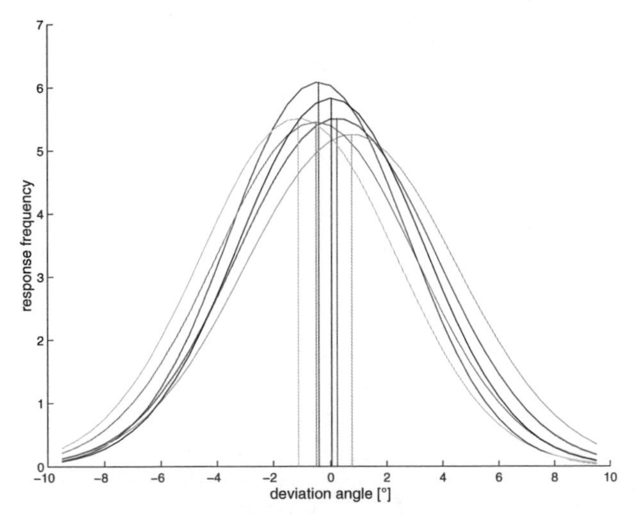

Figure 4.2: The deviation from the predefined path during training is equal to the roll subjects experienced. The different Gaussian fits of the profiles for individual subjects show in addition the small mean variation around the upright position.

The distribution of heading angles during the trials was Gaussian as well. The performed heading turns (relative changes of heading that occurred while following the path) of the subjects correlated with the turns of the virtual path (for all subjects: $p<0.00001$). The correlation over all subjects explains between 43% and 95% of the variability (r^2: 0.84, 0.87, 0.43, 0.83, 0.83, 0.95 with mean $r^2=0.83$). In other words, the active group actually performed the turns asked for. See also Figure 4.3 for the linear fit for one individual, representative subject.

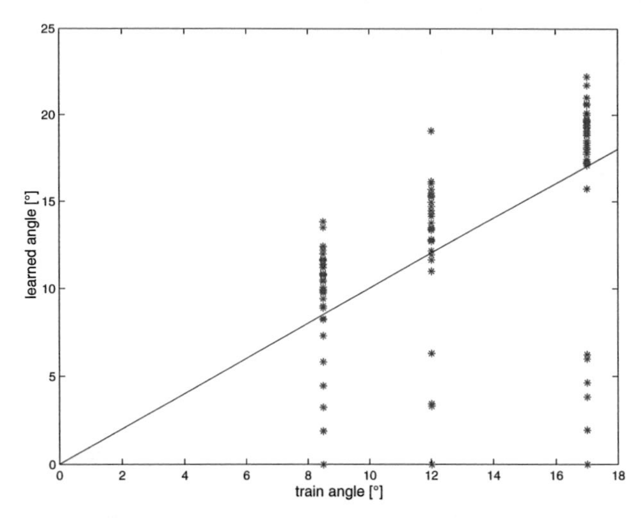

Figure 4.3: This plot shows the correlation of a subjects turns during the training and the underlying turns of the predefined paths. The data points that do not lay close to the diagonal were from trails, where the subject did not stay within the required 10° of maximum roll during the turn. These trials ended at oblique turn angles.

4.2.2 Testing phase

From the active group one subject's data was eliminated due to the subject's inability to consistently perform the task on the previously reached performance level. For the remaining seven subjects, the overall correlation between learned and reproduced angles proved to be significant (t(25)=4.50, p=0.00007***). See Figure 4.4.a for the overall correlation. Individually, two of the subjects showed only weak correlations between learned angles and their reproductions, all other subjects showed highly significant correlations between learned and reproduced angles ($p<0.005$). However, the individual variability was rather high and the overall correlation explains only 12% of the variability (r^2=0.12).

In the passive group, subjects also learned the angles turned and were in general able to reproduce correlated turn amplitudes. One subject had to be excluded from the analysis, since the data showed that this subject obviously did not understand the task. For the remaining six subjects, the overall correlation between learned and reproduced angles proved to be highly significant (t(25)=4.06, p=0.0002***, r^2=0.40). Figure 4.4.b shows the reproduced angles plotted against the angles subjects learned. In comparison to the active group, it has to be noted that the mean value for the angles is smaller. This might be due to the fact that subjects in the active condition overshot the turns by riding on the outside of the curve and had therefore coded a bigger angle. An additional explanation for the increased variability of the active group would be that the active group was preoccupied by the balance task and did not fully concentrate on the turn amplitudes.

Figure 4.5 shows the overall performance of the active and passive groups in terms of the response distribution of the reproduced angle. The distributions overlap to a large degree, but since the chosen angles were so close to each other (8.5°, 12.0°, 17.0°) one could have

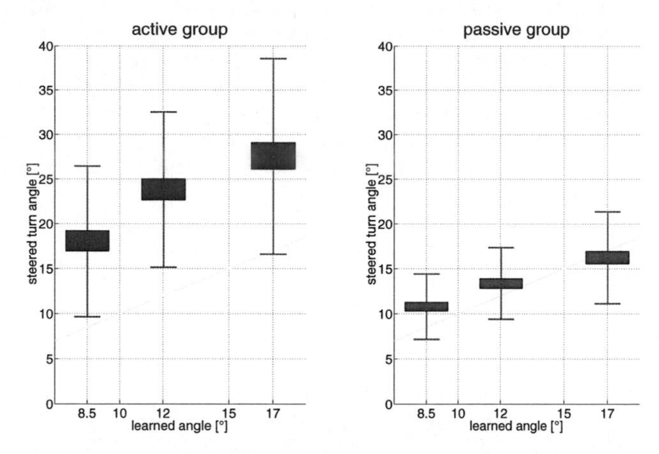

Figure 4.4: The correlation of learned and reproduced angles shows that the subjects indeed reproduced turn amplitudes from memory. There seems to be no qualitative difference between the active and passive group. The bars refer to the standard error of the mean, with "whiskers" depict one standard deviation.

expected this. In addition, the large variability between subjects spread the fitted Gaussian curves. This type of distribution allowed us to perform an ANOVA on the data. The mean values proved to be significantly different (see next paragraph for ANOVA analysis).

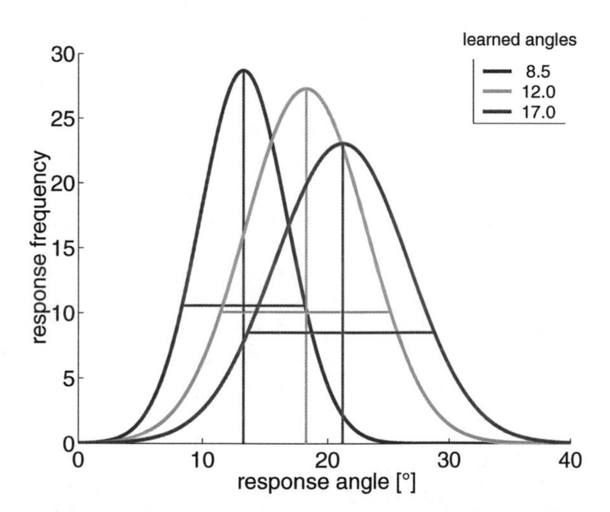

Figure 4.5: The graph shows the response frequency of the reproduced angles grouped for the actually learned angle. The distributions were Gaussian shaped but have a high overlap. Nonetheless, the mean values do significantly differ from each other as confirmed by the ANOVA (see text and Table 4.2).

One of the subjects in the active group systematically "reproduced" too many turns for some of the factorial combinations and was therefore excluded from further analysis. For the remaining 12 subjects (six subjects from each group), a four factorial ANOVA (active/passive x visual gain x vestibular gain x learned angle) was performed. Due to the design of the experiment the factor active/passive was varied between the groups; the other factors' conditions were varied within subjects. The ANOVA revealed significant effects for all four varied conditions and two interactions. Table 4.2 summarizes the output of the analysis based on a α-Value of 5%.

varied factor	F-value and p-value	significance level
active/passive condition	$F(1,10)=11.7$, $p=0.007$	**
visual gain factor	$F(2,20)=9.03$, $p=0.002$	**
vestibular gain factor	$F(2,20)=20.3$, $p<0.001$	***
learned angle	$F(2,20)=52.5$, $p<0.001$	***
visual gain – vestibular gain	$F(4,40)= 3.48$, $p=0.016$	*
learned angle – active/passive	$F(2,20)= 4.28$, $p=0.028$	*
all other interactions	$p>0.10$	(n.s.)

Table 4.2: This table summarize the results of the four factorial ANOVA. All varied factors proved to be significant. In addition, two interactions showed significant effects, the visual gain and vestibular gain factor as well as the learned angle and the active/passive condition.

In Figure 4.6, the interaction between visual and vestibular gain factor influences is depicted for the two groups separately. In this figure the data is scaled by the ration of individual mean of the subjects and the overall mean. This scaling eliminates the inter individual differences of the subjects, but keeps the general pattern of all other effects. Both panels plot the steered angle against the visual gain factor. The vestibular gain factor is coded by the color of the lines. The turned angle in this plot is equal for the visual and vestibular modality only for those factor combinations where the gain factors are equal. In all gain factor combinations the turned angle for one specific modality is the product of the specific gain factor and the steered angle. For example, for the active group in the combination of vestibular gain=$\sqrt{2}$ and visual gain=$\frac{1}{\sqrt{2}}$ the mean steered angle was about 20°. This results is a visually turned angle of roughly 14°, whereas the vestibular turned angle was twice as much (approx. 28°). By the experimental design, the angle plotted here is not equal to one or the other modality, but refers to an internally calculated angle which is correlated to the actual joystick response of the subject. The main difference between left and right panel correspond to the higher steering angles of the active group (compare with Fig. 4.4). Nonetheless, the overall interaction of visual and vestibular gain factor can be seen in both panels: Changing one or the other gain does not, independently from the other gain factor, result in an effect. The effects of visual and vestibular gain factors do not simply result in a linear sum of both effects. Interestingly, the graphs show the smallest variation in the conditions where one of the gain factors (either visual or vestibular gain) is maximal. The respective other gain factor does not vary the response as much as in the other conditions.

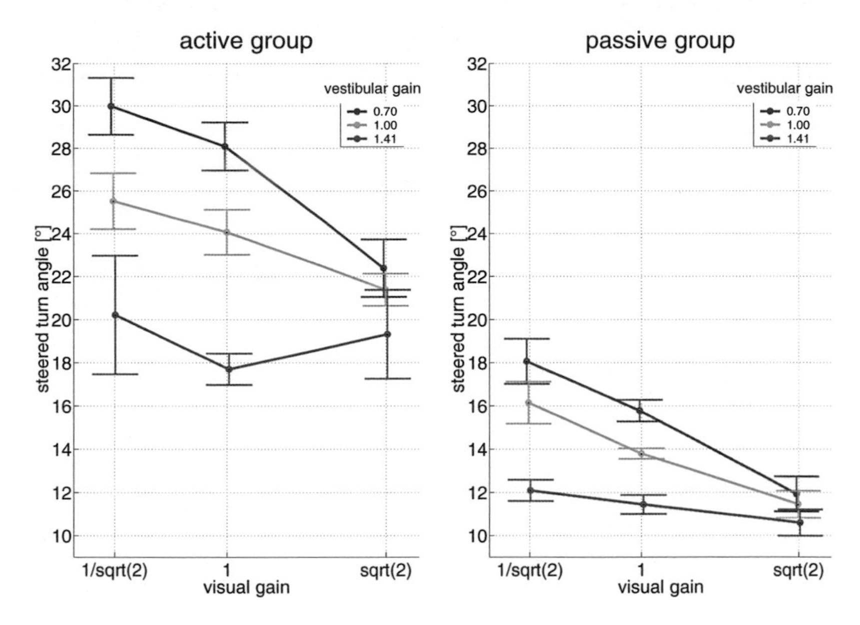

(a) Subjects from the active group reproduced much larger angles.

(b) Subjects from the passive group were more accurate, but showed the same pattern.

Figure 4.6: Both graphs display the interaction between the influence of the visual and the vestibular gain factor on the reproduced angle. The graphs show all data collapsed across all angles. Inter-individual differences were eliminated by scaling the data with the ratio of individual mean and overall mean. The "whiskers" depict the standard error of the mean.

subject	**roll**	**roll+yaw**	**vest + vision**
YSK	2	1	2
QHF	1	1	2
NSL	1	1	1
UCY	1	1	1
JVA	6	3	4
OTI	1	1	2
mean	2	1.33	2
std	2	0.82	1.10

Table 4.3: The table shows the performance active subjects reached during training.

4.3 Discussion

In general, subjects of the active group were able to learn the balance task and use this ability for the test phase of the experiment. They were able, even though they had to concentrate on the balance task, to memorize the sequence of turns. Despite the high within and between subject variability, 13 of the 15 subjects were, overall, able to perform the task of the test phase in a meaningfull way. Subjects learned and memorized curves of the virtual path and were able to reproduce the amplitudes of the turns. The variation of visual and vestibular gain factors had a major influence on the reproduced angles: Subjects compensated for the changes in gain, but due to the conflict condition imposed on them, they chose for a given combination of visual and vestibular gain a compromise between the two modalities. Nonetheless, the modality with the bigger gain factor had dominate influence. Especially for the conditions where either one of the modalities was amplified and maximal, the other modality had a reduced influence. This results can be interpreted as a maximum rule for cue integration.

4.3.1 Answers to the main questions

Coming back to the questions formulated at the begin of this chapter, we can summarize the answers based on the results as follows:

Can humans maintain balance without vision?

Yes, they can. Our subjects could maintain balance and in addition compensate for additional external disturbances. The subjects reached comparable performance levels for the balance task. Subjects could even hold their balance solely based on the vestibular input, since the simulated visual horizon did not show any roll rotations. Comparing the precision reached for the condition with and without a visual stimulus, one has to conclude that the visual stimulus was at least not interfering with the balancing task even though it did not provide any helpful information.

Is it possible to ignore additional rotations which are not related to the task?

Yes, it is. More specifically, subjects were able to adapt to the additional rotation in less time than it had taken them to learn the main balancing task. Nonetheless, subjects reported that the additional rotation was disturbing in the beginning, but it felt more realistic (especially when the visual stimulation was provided) having the heading turns matching the optic flow. Subjects showed that the additional rotation did not interfere by reaching at least the same level of performance when training with heading changes.

Which heading information is coded during path following?

The results allow the interpretation of separated storage of both modalities, meaning that visual turned angles are stored somewhere other than vestibular turned angles. If this proposal is accepted, one has to conclude that during the reproduction of the angle, no cross-modal check occurred which would have pointed out the mismatch of either one of the modalities. The gain factor between visual and vestibular modality was varied between 0.5 and 2.0, changing the modality specific angle over the complete range of the used heading changes. Since subjects were able to distinguish between these small angles, the mismatch between the modalities should have been noticed by most of the subjects. Since this was not the case, we are in favour of the following interpretation.

The opposite interpretation of a joint storage of heading changes would not result in these kind of implications. Moreover, a cue integration model can be assumed using the same storage for the joint percept. One can think here of a distributed representation in the sense of the second principle defined by O'Reilly (1998). If the percept of both modalities was stored in one representation in the coherent condition during the learning phase, then there is only this one representation to compare to the percept during reproduction phase. Having a conflicting gain ratio between visual and vestibular modality during reproduction now allows the subject to turn until one of both modalities reaches the required turn amplitude. Not denying the impact of the other modality, the influence could manifest itself in the undershot of the mean angle turned in comparison to the learned condition. In other words, the modality with the smaller gain factor has less impact and the response is dominated by the modality with the bigger gain factor. This interpretation is called a "max-rule" concept of cue integration.

Is the information differently coded when being moved passively?

Our results can not confirm any hypothesis that in an active condition the coded heading change should be more precise. Quite to the contrary, subjects in the passive condition had performed better. In the active condition the reproduced turns were on average 91.1% bigger than the learned turning amplitudes. In contrast, the subjects in the passive condition overshot only by 11.2%. Additionally, there was a small tendency of the passive group to have less variance in the reproduced angles. Both effects can, as earlier pointed out, be the result of the active subjects riding on the outside of the curves actually using bigger turn amplitudes in the coding phase. In addition, the coding might be influenced by the balance task binding some of the subjects mental activity. This would confirm results from Yardley and Higgins (1998) for an experiment where mental activity interfered with the perception of self-rotation. The spatial updating (in this experiment without vision) required active monitoring and thereby was disturbed by the additional mental load. In contrast, Jürgens et al. (1999) found active turns were judged and executed more precisely than passive turns. However, there is no basis for concluding different models for coding the information.

4.3.2 Comparison with results from the literature

Other studies have concluded that the vestibular system does appear not to be necessary to perform path integration. More specifically, Warren and Hannon (1988) pointed out that the direction of self motion can be derived from optic flow alone. Glasauer et al. (1994) compared labyrinthine-defective subjects with normals in a task where subjects had to walk towards a previously seen target. Taking both results, one has to conclude that without the vestibular sense navigation even in VR environments should be possible, as Riecke et al. (2000) proved for triangle completion tasks. Nonetheless, the latter study pointed out that there was probably no spatial updating necessary for the solution of the task. Moreover, Riecke et al. reported a correlation of mental spatial ability with the performance in the experiments. To test spatial updating as opposed to cognitive strategies to solve spatial tasks, one has to compare more natural tasks in the future involving a stimulation of multiple or all senses.

Understanding such experiments demands a deeper consideration of models for cue integration. Since in our experiments only one subject spontaneously reported a conflict between vestibularly and visually perceived angle, the integration of both cues seems to be very flexible. This confirms earlier findings from Ivanenko et al. (1998). They had shown that the visual system can recalibrate the vestibular system, suggesting a visual dominance in visual/vestibular integration. The use of vestibular training in the present experiment demonstrates the relevance of the vestibular signal in our task. We suggest a dominance, at least for the comparison of a percept with a memorized one, for the modality with the bigger signal. In contrast, a conceptual model of high-threshold value summation (vestibular (head in space), neck proprioception (trunk relative to body) and leg proprioception (leg relative to trunk)) was already proposed by Mergner, Hlavacka, and Schweigart (1993) and later discussed in the context of microgravity (Mergner and Rosemeier, 1998). Mergner and Rosemeier (1998) point out neural structures which support the idea of a linear summation of somatosensoric and vestibular signals.

Chapter 5

Summary

Over the course of evolution, humans as well as other animals learned to navigate through complex environments mainly for two goals: to find food and to find the way back to shelter. Therefore, it is important for most moving organisms to know their location in the world and maintain some internal representation of it. For higher species it is most likely that multiple sensory systems provide information to solve this task.

This thesis discussed the above task in the light of the concept of spatial reference frames. Most sensory modalities are known to provide some information about position or movements in space. Humans primarily use sight, vestibular input, proprioception, motor commands, and auditory input for determining self position and movements. Taste and smell are not believed to have a major influence if the other senses are well functioning. Here, we make the assumption that the information provided from different senses has to be integrated at some point to establish a representation of one's position and motion in relation to the world. In order to integrate modality specific spatial information, it has to be transformed and then integrated into the global representation. Since the representation of modality specific perception in the overall frame of reference is assumed to be different from the representation of perception at the sensory level, a transformation process must take place in between. We further assume that the existence of a global representation is the basis for the usual percept of being only at one location at a time. The belief in one unique frame of reference therefore depends on the percept of individual senses and the interaction with a somehow memorized status. Naturally, the overall frame of reference is sometimes fooled and we misperceive our movements or position in space. This situation often occurs when the different sensory modalities do not agree in their perception of the motion or stability of the body in space. For the sensory integration we propose a dynamic process which, in addition to the pure information from the sensory system, takes the probability of that specific perception into account. This approach allows an appropriate representation to be constructed even in the presence of noisy or unusual perception, or even malfunction of the sensory system. Recently, a study by Triesch, Ballard, and Jacobs (2000) presented data within the visual domain which are consistent with the fast dynamic update of the weights in a neural model based on the reliability of different object attributes.

Two complementary methods are used to study the integration process. The first describes the perception of single, isolated sensory modalities in order to state their contribution. The other regards the whole system and changes small parts of one modality and specifies the change of overall performance. Both methods change one modality at a time, but differ in the relevance of the unchanged modalities. Implementing both is possible using synthetic stimuli in Virtual Reality. The simulated version of the world can provide modality specific input to the human sensory system with high reliability, accuracy and full control of stimulus features.

Therefore, this thesis introduced a laboratory (Motion-Lab) which provided Virtual Reality applications involving simulations for all the relevant senses. Both research methods (single and multiple modalities) were demonstrated in two experiments about the ability of humans to perform spatial updating.

The first experiment asked which information humans use to orient in the environment and maintain an internal representation about the current location in space. The underlying question in this experiment focused on the general dimension of the percept. Specifically, is the distance, velocity or acceleration directly perceived or does one derive estimates of those values? The sensor organs (canals and otoliths of the vestibular system) are certainly stimulated by angular and linear accelerations, respectively. But are those accelerations transformed, mathematically integrated, into a velocity estimate? Further, is the velocity value, if it exists, usable for integrating a second time to come up with an estimate for traveled distance or angle?

In order to perceive our environment as stable during movements, we have to stabilize ourselves, too. The second experiment posed the question of whether we can stabilize ourself in space and learn certain characteristics of a path. Specifically, can we code the angular amplitude (heading turns) in space during a task where we follow a virtual path without actually seeing it? Does the path following allow us to learn the path and repeat the turns we learned? In the experiment, we focused on two cues that provide no absolute spatial reference: optic flow and vestibular cues. Specifically, we asked whether both visual and vestibular information are stored and can be reproduced later. The experiment therefore tried to separate which information (visual or vestibular) is used to reproduce the memorized path. Further, are those modalities integrated into one coherent percept or is memory modality specific?

Both experiments are connected by the question of how we perceive turns. In the first experiment, verbal judgements about the heading changes were compared with linear movements. In the second experiment, the turns were presented in multiple modalities and were tested in a cue conflict condition.

The results of both experiments can be summarized as follows. In the first experiment, 12 blindfolded subjects gave verbal judgements of their distance traveled, maximum velocity and peak acceleration for short vestibular stimuli. The stimuli were designed to independently vary the distance and peak acceleration without allowing correlation to movement time. The judgements were highly correlated to the physical properties of the presented stimuli, demonstrating that subjects were able to perform the task. Nonetheless, the acceleration estimates were, for most of the subjects, very similar to the velocity judgements and highly correlated to the peak velocity. One possible explanation is that since velocity at least for angular movements is encoded at a very early stage in the sensor organ, the acceleration signal may not be available for later cognitive access. Interestingly, the judgements for linear movements in the horizontal plane were not different from angular movements around the body vertical, regardless of the different sensor organs responsible for the perception of angular and linear acceleration. One interpretation of this fact is that verbal judgements access the overall reference frame which has common features for position and orientation representation. In sum, the vestibular system provides a good spatial reference frame which enables spatial updating.

In the second experiment, the relative contribution of multiple sensory systems was examined for the same spatial updating process. Two groups (active/passive) learned the geometry of a path based on optic flow and whole body turns round the body vertical. The passive group was driven along the virtual path and the active group had to perform a vestibular balance task in order to follow the same path. Despite the high variability within and between subjects, 13 of 15 subjects were, overall, able to perform the task in a meaningful way. Subjects learned and memorized the curves of the virtual path and were able to reproduce the amplitudes of the turns. During the reproduction, the ratio of the stimulus magnitude for the visual and vestibular system was varied. The variation of visual and vestibular gain factors had major influence on the reproduced angles: Subjects compensated for the changes in gain, but due to the conflict condition imposed on them, they chose a compromise between the two modalities for a given combination of visual and vestibular gain. Nonetheless, the modality with the bigger gain factor had the dominant influence. Moreover, a cue integration model can be assumed using the same storage for a joint percept. One can think here of a distributed representation of orientation in space. If the percept of both modalities was stored in one representation in the coherent condition during the learning phase of the path, then there would only be this representation to compare during the reproduction phase. Having a conflicting gain ratio between visual and vestibular modality during reproduction allows the subjects to turn until one of both modalities reaches the required turn amplitude. The impact of the other modality in the joint representation was expressed by an undershoot of the mean angle turned in comparison to the learned condition. The other modality had reduced influence, especially, for the conditions where either one of the modalities was amplified and maximal. These results can be interpreted as a "max-rule" for cue integration. In other words, the modality with the smaller gain factor has less impact and the response is dominated by the modality with the bigger gain factor. The above can be used to propose a model of cue integration where a dynamically weighted sum of all modalities is integrated in order to come up with a coherent percept and memory.

In order to study human behavior in a complex environment, it is important that the experimenter has full control over the stimulus. These studies were conducted in the Motion-Lab which is a Virtual Reality laboratory combining simulations for multiple senses. Beside the experiments described above, the design and construction of the lab as well as the conception and implementation of the necessary software is part of this thesis.

In the Motion-Lab, it is possible to stimulate four senses at the same time: vision, acoustics, touch, and the vestibular sense of the inner ear. Special purpose equipment is controlled by individual computers to guarantee optimal performance of the modality specific simulations. Local loops implement modality specific feedback. The coupling of the modality specific simulations is done in a client/server architecture by connecting all computers to a central VR simulation. Due to frequency differences in the simulations imposed by the natural latency and time resolutions of different senses, the data transmission between clients and servers is realized as asynchronous communication. The transfered data is coded in a way that the loss of single packages is acceptable. Where needed, interpolation and extrapolation methods are used to smooth the data stream. For the programmer, the distributed components are accessible through a library for multiple operating systems[1]. The necessary communication between client and server is transparent for the

[1]Currently Windows95, WindowsNT, IRIX, and Linux.

programmer, letting all devices appear to be connected locally. A unified devices layer which is implemented by the devices' servers enables easy exchange of the devices' components and computers.

A large variety of equipment is used in the Motion-Lab for the realization of the VR simulation. Several input devices can be used to allow the user to interact with the simulated world. Joysticks, a force feedback steering wheel, and a six degree of freedom tracking device are available. Immediate feedback is provided by the output devices for multiple modalities. For the presentation of visual stimuli several graphics libraries can be used independently from the actual simulation. The rendered virtual scenery is presented in stereo via a head mounted display. Coupled to a tracking device for head movements, the virtual camera can be controlled appropriately. Even when the observer does not move actively, he can be moved in space by means of a motion platform. The platform can perform movements in all six degrees of freedom independently and deliver high accelerations. Acoustic simulations include synthetic speech and other stereo sound effects presented via special noise canceling headphones. Vibrations can be simulated either by the platform or for higher frequencies by force transducers. Since the physical movement of the platform can not extend beyond a certain range, vibrations are used to suggest velocity coupled cues to the observer. All these devices are covered within the client/server architecture by individual devices servers.

The distributed VR approach of the Motion-Lab has several advantages over the classic mainframe based VR system. The general architecture enables easy extension of the lab and will keep the development moving on in the future. Decoupled simulations for the different modalities were demonstrated as security features. Coupling between frequency demands of different modalities was achieved by inter- and extrapolation methods in combination with the communication model. Real parallel execution is supported and enhanced by the asynchronous communication. The overall performance of the distributed system is sufficient for high level VR applications.

Nonetheless, several questions were not addressed in this thesis. It remains unclear whether the motion platform can deliver a natural feeling of changed location especially in the context of spatial updating for large scale movements. In particular, in the light of the results of the present experiments it appears that the perception of self position and self motion based on the vestibular system would be sufficient at least to disagree with the other senses in the integration process. Ways to trigger the reliability of the perception of different senses in the proposed model are yet to be explored in order to confirm these speculations. In general, more experiments, which focus on the transfer of results gained in VR, are needed. The study of high level behavior like navigation or spatial updating in VR experiments is often subject to criticism emphasizing the dominant use of cognitive strategies in contrast to pure psychophysical experiments. Nonetheless, more experiments are necessary to investigate the role the different modalities play in the spatial updating mechanisms. On the other hand, the effects of the spatial update on the perception of different modalities is still subject to examination. Whether, for example, the change in position and orientation can generate expectancies and thereby enhance visual perception remains unclear.

References

Arthur, K. W. (2000). *Effects of Field of View on Performance with Head-Mounted Displays*. Ph.D. thesis, Department of Computer Science, University of North Carolina, Chapel Hill. [Online] Available: http://www.cs.unc.edu/ arthur/diss/.

Atkins, J. E., Fiser, J., and Jacobs, R. A. (2000). Experience-dependent visual cue integration based on consistencies between visual and haptic percepts.. (in press).

Bakker, N. H., Werkhoven, P. J., and Passenier, P. O. (1999). The effects of proprioceptive and visual feedback on geographical orientation in virtual environments. *Presence: Teleoperators & Virtual Environments*, **8**(1), 36 – 53.

Bangay, S., Gain, J., Watkins, G., and Watkins, K. (1997). Building the second generation of parallel/distributed virtual reality systems. *Parallel Computing*, **23**(7), 991–1000.

Bayliss, J. D., and Ballard, D. H. (2000). A virtual reality testbed for brain-computer interface research. *IEEE Trans. Rehabil. Eng.*, **8**(2), 188 – 190.

Beall, A. C., and Loomis, J. M. (1997). Optic flow and visual analysis of the base-to-final turn. *Int. J. Aviat. Psychol.*, **7**(3), 201 – 223.

Begault, D. R. (1994). *3-D sound for virtual reality and multimedia*. Boston: Academic Press Professional.

Berthoz, A., Israël, I., Georgesfrancois, P., Grasso, R., and Tsuzuku, T. (1995). Spatial memory of body linear displacement: What is being stored?. *Science*, **269**(5220), 95–98.

Bülthoff, H. H., Foese-Mallot, B. M., and Mallot, H. A. (1997). Virtuelle Realität als Methode der modernen Hirnforschung. In T. W. H. Krapp (Ed.), *Künstliche Paradiese, Virtuelle Realitäten. Künstliche Räume in Literatur-, Sozial- und Naturwissenschaften* (pp. 241–260). München: Wilhelm Fink Verlag.

Bülthoff, H. H., Riecke, B. E., and van Veen, H. A. H. C. (2000). Do we really need vestibular and proprioceptive cues for homing. *Invest. Ophthalmol. Vis. Sci.*, **41**(4), 225B225.

Bülthoff, H. H., and van Veen, H. A. H. C. (1999). Vision and Action in Virtual Environments: Modern Psychophysics in Spatial Cognition Research. Tech. Rep. 77, Max-Planck-Institute for Biological Cybernetics, Tübingen, Germany.

Burdea, G. (1993). Virtual Reality Systems and Applications. In *Electro'93 International Conference*, Short Course, p. 164 pp. Edison, NJ.

Burdea, G., and Coiffet, P. (1994). *Virtual Reality Technology*. New York: John Wiley & Sons. Inc.

Chance, S. S., Gaunet, F., Beall, A. C., and Loomis, J. M. (1998). Locomotion mode affects the updating of objects encountered during travel: The contribution of vestibular and proprioceptive inputs to path integration. *Presence: Teleoperators & Virtual Environments*, **7**(2), 168–178.

Chatziastros, A., Wallis, G. M., and Bülthoff, H. H. (1997). The influence of road markings and texture on steering accuracy in a driving simulator. *Invest. Ophthalmol. Vis. Sci.*, **38**(4), 383 – 383.

Cheng, D. Y. (1993). A Survey of Parallel Programming Languages and Tools. Tech. Rep., NASA Ames Research Center, Moffett Field, CA 94035.

Cobb, S. V. G., Nichols, S., Ramsey, A., and Wilson, J. R. (1999). Virtual reality-induced symptoms and effects (VRISE). *Presence: Teleoperators & Virtual Environments*, **8**(2), 169 – 186.

Coulouris, G., Dollimore, J., and Kindberg, T. (1994). *Distributed Systems - Concepts and Design* (second Edition). Wokingham, England: Addison-Wesley Publishing Company.

Crowell, J. A., Banks, M. S., Shenoy, K. V., and Andersen, R. A. (1998). Visual self-motion perception during head turns. *Nat. Neurosci.*, **1**(8), 732 – 737.

Cunningham, D. W., von der Heyde, M., and Bülthoff, H. H. (2000a). Learning to drive with delayed visual feedback. *Invest. Ophthalmol. Vis. Sci.*, **41**(4), S48.

Cunningham, D. W., von der Heyde, M., and Bülthoff, H. H. (2000b). Learning to drive with delayed visual feedback. In H. Bülthoff, M. Fahle, K. Gegenfurtner, and H. Mallot (Eds.), *Beiträge der 3. Tübinger Wahrnehmungskonferenz*, p. 164 Max-Planck-Institute for Biological Cybernetics, Germany. Knirsch Verlag, Kirchentellinsfurt, Germany.

Dell'Osso, L. F., and Daroff, R. B. (1990). Eye movement characteristics and recording techniques. In J. S. Glaser (Ed.), *Neuro-Ophthalmology* (2 Edition).Kap. 9, (pp. 279 – 297). Philadelphia, PA: J. B. Lippincott Company.

Demuynck, K., Broeckhove, J., and Arickx, F. (1998). The VEplatform system: A system for distributed virtual reality. *Future Generation Computer Systems*, **14**(3-4), 193 – 198.

Distler, H. K., van Veen, H. A. H. C., Braun, S., and Bülthoff, H. H. (1997). Untersuchung Wahrnehmungs- und Verhaltensleistungen des Menschen in virtuellen Welten. In W. Bruns, E. Honecker, B. Robben, and I. Rügge (Eds.), *Workshop 'Vom Bildschirm zum Handrad: Computer(be)nutzung nach der Desktop-Methapher'*. Forschungszentrum Arbeit und Technik, Universität Bremen, Germany, artec-paper 59 february 1998.

Distler, H. K. (2000). *Wahrnehmung und virtuelle Welten*. Ph.D. thesis, Fakultät Biologie, Eberhard-Karls-Universität Tübingen. in publication.

Eliasson, A.-C., Rösblad, B., and Häger-Ross, C. (2000). Effects of Practice with Computer Games on the Control of Planar Reaching Movements in 6-year Old Prematurely Born Children. In *International Meeting on "Therapeutical Interventions in Motor Disorders: Neural Mechanisms and Clinical Efficacy"* Groningen, Holland.

Ernst, M. O., Banks, M. S., and Bülthoff, H. H. (2000). Touch can change visual slant perception. *Nat. Neurosci.*, **3**(1), 69 – 73.

Fechner, G. T. (1860). *Elemente der Psychophysik*, Vol. 1. Leipzig: Breitenkopf und Harterl.

Fichter, E. F. (1986). A Stewart Platform-Based Manipulator: General Theory and Practical Construction. *The International Journal of Robotics Research*, **5**(2), 157 – 182.

Frécon, E., and Stenius, M. (1998). Dive: a scaleable network architecture for distributed virtual environments. *Distributed Systems Engineering*, **5**(3), 91 – 100.

Geiger, S., Gillner, S., and Mallot, H. A. (1997). Global versus local cues for route finding in virtual environments. *Perception*, **26**(Suppl.), 56b.

Geist, A., Beguelin, A., Dongarra, J., Jiang, W., Manchek, R., and Sunderam, V. (1994). *PVM: Parallel Virtual Machine - A User's guide and Tutorial for Networked parallel Computing*. Cambridge, London: MIT Press.

Gilkey, R., and Weisenberger, J. (1995). The sense of presence for the suddenly deafened adult: Implications for virtual environments. *Presence: Teleoperators & Virtual Environments*, **4**(4), 357 – 363.

Gillner, S., and Mallot, H. A. (1996). Place–based versus view–based navigation: Experiments in changing virtual environments. *Perception*, **25**(Suppl.), 93.

Glasauer, S., Amorim, M. A., Vitte, E., and Berthoz, A. (1994). Goal-directed linear locomotion in normal and labyrinthine- defective subjects. *Exp. Brain Res.*, **98**(2), 323 – 335.

Greenhalgh, C. (1998). Awareness-based communication management in the MASSIVE systems. *Distributed Systems Engineering*, **5**(3), 129 – 137.

Haller, H. (1999). Was bedeutet Signifikanz? Eine empirische Semesterarbeit. MPI für Bildungsforschung, Berlin.

Harada, H., Kawaguchi, N., Iwakawa, A., Matsui, K., and Ohno, T. (1998). Space-sharing architecture for a three-dimensional virtual community. *Distributed Systems Engineering*, **5**(3), 101 – 106.

Harris, L. R., Jenkin, M., and Zikovitz, D. C. (1998). Vestibular cues and virtual environments. In *IEEE Virtual Reality Annual International Symposium (VRAIS)*, pp. 133 – 138 Atlanta, GA.

Harris, L. R., Jenkin, M., and Zikovitz, D. C. (1999). Vestibular cues and virtual environments: choosing the magnitude of the vestibular cue. In *1st IEEE International Conference on Virtual Reality*.

Heilig, M. (1960). Stereoscopic-Television Apparatus for Individual Use. US Patent No. 2,955,156.

Hendrix, C., and Barfield, W. (1996). The sense of presence within auditory virtual environments. *Presence: Teleoperators & Virtual Environments*, **5**(3), 290–301.

Hlavacka, F., Mergner, T., and Bolha, B. (1996). Human self-motion perception during translatory vestibular and proprioceptive stimulation. *Neurosci. Lett.*, **210**(2), 83 – 86.

Horn, C. (1989). Is Object Orientation a Good Thing for Distributed Systems?. In W. Schröder-Preikschat and W. Zimmer (Eds.), *Progress in Distributed Operating Systems and Distributed System Management*, Vol. 433 of *Lecture Notes in Computer Science*, pp. 60 – 74 Berlin – Heidelberg – New York. Springer-Verlag.

Hornby, A. S. (Ed.). (1983). *Oxford Advanced Learner's Dictionary od Current English* (14 Edition). Bielefeld: Cornelsen-Velhagen & Klasing.

Husemann, D. (1996). *Multimedia Data Streams in Distributed Object-Oriented Operating Systems*. Ph.D. thesis, Institut für Mathematische Maschinen und Datenverarbeitung (Informatik), Friedrich-Alexander-Universität Erlangen-Nürnberg, Germany.

Israël, I., Sievering, D., and Koenig, E. (1995). Self-rotation estimate about the vertical axis. *Acta Oto-Laryngol.*, **115**(1), 3 – 8.

Ivanenko, Y., Grasso, R., Israël, I., and Berthoz, A. (1997a). Spatial orientation in humans: perception of angular whole-body displacements in two-dimensional trajectories. *Exp. Brain Res.*, **117**(3), 419 – 427.

Ivanenko, Y. P., Grasso, R., Israël, I., and Berthoz, A. (1997b). The contribution of otoliths and semicircular canals to the perception of two-dimensional passive whole-body motion in humans. *J. Physiol.-London*, **502**(1), 223 – 233.

Ivanenko, Y. P., Viaud-Delmon, I., Siegler, I., Israël, I., and Berthoz, A. (1998). The vestibulo-ocular reflex and angular displacement perception in darkness in humans: adaptation to a virtual environment. *Neurosci. Lett.*, **241**(2-3), 167 – 170.

Jones, T. A., Jones, S. M., and Colbert, S. (1998). The adequate stimulus for avian short latency vestibular responses to linear translation. *J. Vestib. Res.-Equilib. Orientat.*, **8**(3), 253 – 272.

Jürgens, R., Boß, T., and Becker, W. (1999). Estimation of self-turning in the dark: comparison between active and passive rotation. *Exp. Brain Res.*, **128**(4), 491 – 504.

Jungclaus, N. (1998). *Integration verteilter Systeme zur Mensch-Maschine-Kommunikation*. Ph.D. thesis, Universität Bielefeld – Technische Fakultät.

Köhler, W., Schachtel, G., and Voleske, P. (1995). *Biostatistik* (2 Edition). Berlin - Heidelberg: Springer.

King, R., and Oldfield, S. (1997). The impact of signal bandwidth on auditory localization: Implications for the design of three-dimensional audio displays. *HUMAN-FACTORS*, **39**(2), 287 – 295.

Klaeren, H. A. (1991). *Vom Problem zum Programm: eine Einführung in die Informatik* (2 Edition). Stuttgart: Teubner.

Klatzky, R. L., Loomis, J. M., Beall, A. C., Chance, S. S., and Golledge, R. G. (1998). Spatial updating of self-position and orientation during real, imagined, and virtual locomotion. *Psychol. Sci.*, **9**(4), 293 – 298.

Kolev, O., Mergner, T., Kimmig, H., and Becker, W. (1996). Detection thresholds for object motion and self-motion during vestibular and visuo-oculomotor stimulation. *Brain Res. Bull.*, **40**(5-6), 451 – 457.

Kowal, K. H. (1993). The range effect as a function of stimulus set, presence of a standard, and modulus. *Percept. Psychophys.*, **54**(4), 555 – 561.

Lampson, B. W., Paul, M., and Siegert, H. J. (Eds.). (1981). *Distributed Systems - Architecture and Implementation*, Vol. 105 of *Lecture Notes in Computer Science*. Berlin – Heidelberg – New York: Springer-Verlag.

Latoschik, M. E., and Wachsmuth, I. (1998). Exploiting Distant Pointing Gestures for Object Selection in a Virtual Environment. In *Gesture and Sign Language in Human-Computer Interaction*, Lecture Notes in Computer Science, pp. 185–196 Berlin – Heidelberg. Springer-Verlag Berlin.

Loomis, J. M., Blascovich, J. J., and Beall, A. C. (1999). Immersive virtual environment technology as a basic research tool in psychology. *Behav. Res. Methods Instr. Comput.*, **31**(4), 557 – 564.

Merfeld, D. M., Zupan, L., and Peterka, R. J. (1999). Humans use internal models to estimate gravity and linear acceleration. *Nature*, **398**(6728), 615 – 618.

Mergner, T., Hlavacka, F., and Schweigart, G. (1993). Interaction of vestibular and proprioceptive inputs. *Journal of Vestibular Research*, **3**, 41 – 57.

Mergner, T., Nasios, G., and Anastasopoulos, D. (1998). Vestibular memory-contingent saccades involve somatosensory input from the body support. *Neuroreport*, **9**(7), 1469 – 1473.

Mergner, T., and Rosemeier, T. (1998). Interaction of vestibular, somatosensory and visual signals for postural control and motion perception under terrestrial and microgravity conditions - a conceptual model. *Brain Res. Rev.*, **28**(1-2), 118 – 135.

Mergner, T., Rumberger, A., and Becker, W. (1996). Is perceived angular displacement the time integral of perceived angular velocity?. *Brain Res. Bull.*, **40**(5-6), 467 – 470.

Mergner, T., Siebold, C., Schweigart, G., and Becker, W. (1991). Human perception of horizontal trunk and head rotation in space during vestibular and neck stimulation. *Exp. Brain Res.*, **85**(2), 389 – 404.

Mühlberger, A., Herrmann, M., Pauli, P., Wiedemann, G., and Ellgring, H. (1999). The treatment of flight anxiety by exposure to virtual worlds. *Verhaltenstherapie*, **9**, 51 – 51.

O'Reilly, R. C. (1998). Six principles for biologically based computational models of cortical cognition. *Trends Cogn. Sci.*, **2**(11), 455 – 462.

Paul, M., and Siegert, H. J. (Eds.). (1984). *Distributed Systems - Methods and Tools for Specification*, Vol. 190 of *Lecture Notes in Computer Science*. Berlin – Heidelberg – New York: Springer-Verlag.

Pelz, J. B., Hayhoe, M. M., Ballard, D. H., Shrivastava, A., Bayliss, J. D., and von der Heyde, M. (1999). Development of a Virtual Laboratory for the Study of Complex Human Behavior. *Proceedings of the SPIE - The International Society for Optical Engineering*, **3639**.

Poulton, E. C. (1967). Population norms of top sensory magnitudes and S. S. Stevens' exponents. *Percept. Psychophys.*, **2**, 312 – 316.

Poulton, E. C. (1968). The new psychophysics: Six models for magnitude estimation. *Psychol. Bull.*, **69**, 1 – 19.

Poulton, E. C. (1981a). Human manual control. In V. B. Brooks (Ed.), *Handbook of Physiology*, pp. 1337 – 1389 Bethesda. American Physiological Society.

Poulton, E. C. (1981b). Schooling and the new psychophysics. *Behav. Brain Sci.*, **4**(2), 201 – 203.

Powers, S., Hinds, M., and Morphett, J. (1998). DEE: an architecture for distributed virtual environment gaming. *Distributed Systems Engineering*, **5**(3), 107 – 117.

Reitmayr, G., Carroll, S., Reitemeyer, A., and Wagner, M. G. (1999). DeepMatrix - An open technology based virtual environment system. *Visual Computer*, **15**(7/8), 395 – 412.

Riecke, B. E., van Veen, H. A. H. C., and Bülthoff, H. H. (1999). Is homing by optic flow possible?. *J. Cogn. Neurosci.*, **1**.

Riecke, B. E., van Veen, H. A. H. C., and Bülthoff, H. H. (2000). Visual Homing is possible without Landmaks: A Path Integration Study in Virtual Reality. Tech. Rep. 82, Max-Planck-Institute for Biological Cybernetics, Tübingen, Germany.

Riecke, B. E. (1998). Untersuchung des menschlichen Navigationsverhaltens anhand von Heimfindeexperimenten in virtuellen Umgebungen. Master's thesis, Eberhard-Karls-Universität Tübingen, Fakultät für Physik.

Ryan, M. D., and Sharkey, P. M. (1998). Distortion in Distributed Virtual Environments. In J.-C. Heudin (Ed.), *Virtual Worlds 98*, Vol. 1434 of *Lecture Notes in Computer Science*, pp. 42 – 48 Berlin – Heidelberg – New York. Springer-Verlag.

Salgian, G. (1998). *Tactical Driving using Visual Routines*. Ph.D. thesis, Computer Science Dept., University of Rochester.

Sellen, K. (1998). Richtungsschätzungen in realen und virtuellen Umgebungen. Master's thesis, Fakultät Biologie, Eberhard-Karls-Universität Tübingen.

Sibigtroth, M. P., and Banks, M. S. (2000). Perspective transformation in heading estimation.. *Invest. Ophthalmol. Vis. Sci.*, **41**(4), 1.

Sowa, T., Fröhlich, M., and Latoschik, M. E. (2000). Temporal Symbolic Integration Applied to a Multimodal System Using Gestures and Speech. In *ISSU 1739*, Lecture Notes in Computer Science, pp. 291 – 302 Berlin – Heidelberg. Springer-Verlag Berlin.

Steck, S., and Mallot, H. (2000). The role of global and local landmarks in virtual environment navigation. *Presence: Teleoperators & Virtual Environments*, **9**(1), 69–83.

Steck, S. D. (2000). *Integration verschiedener Informationsquellen bei der Navigation in virtuellen Umgebungen*. Ph.D. thesis, Fakultät Physik, Eberhard-Karls-Universität Tübingen.

Stevens, S. S. (1957). The direct estimation of sensory magnitudes – Loudness. *Am. J. Psychol.*, **69**, 1 – 25.

Stewart, D. (1965). A platform with six degrees of freedom. *Proc. Inst. Mech. Engr.*, **180**(1), 371 – 386.

Triesch, J., Ballard, D. H., and Jacobs, R. A. (2000). Fast Temporal Dynamics of Visual Cue Integration.. submitted for publication.

van den Berg, A. V. (1999). Predicting the present direction of heading. *Vision Res.*, **39**(21), 3608 – 3620.

van Veen, H. A. H. C., Distler, H. K., Braun, S. J., and Bülthoff, H. H. (1998). Navigating through a virtual city: Using virtual reality technology to study human action and perception. *Future Generations Computer Systems*, **14**(3-4), 231 – 242.

Viirre, E. (1996). Virtual reality and the vestibular apparatus. *IEEE Eng. Med. Biol. Mag.*, **15**(2), 41 – 44.

von der Heyde, M., and Häger-Ross, C. (1998). Psychophysical experiments in a complex virtual environment. In D. J. K. Salisbury and D. M. A. Srinivasan (Eds.), *Proceedings of the Third PHANToM Users Group Workshop, MIT Artificial Intelligence Report No. 1643, MIT R.L.E. TR No.624*, pp. 101 – 104 Cambridge. MIT Press.

von der Heyde, M., Riecke, B. E., Cunningham, D. W., and Bülthoff, H. H. (2000a). Humans can extract distance and velocity from vestibular perceived acceleration. *J. Cogn. Neurosci.*, **1**.

von der Heyde, M., Riecke, B. E., Cunningham, D. W., and Bülthoff, H. H. (2000b). Humans can separately perceive distance, velocity, and acceleration from vestibular stimulation. In H. Bülthoff, M. Fahle, K. Gegenfurtner, and H. Mallot (Eds.), *Beiträge der 3. Tübinger Wahrnehmungskonferenz*, p. 148 Max-Planck-Institute for Biological Cybernetics, Germany. Knirsch Verlag, Kirchentellinsfurt, Germany.

Warren, W. H., and Hannon, D. J. (1988). Direction of self-motion is perceived from optical-flow. *Nature*, **336**(6195), 162 – 163.

Wedlake, M., Li, K. F., and Guibaly, F. E. (1999). The NAVL Distributed Virtual Reality System. In S. Nishio and F. Kishino (Eds.), *AMCP'98*, Vol. 1554 of *Lecture Notes in Computer Science*, pp. 177 – 193 Berlin – Heidelberg. Springer.

Wiest, W. M., and Bell, B. (1985). Stevens exponent for psychophysical scaling of perceived, remembered, and inferred distance. *Psychol. Bull.*, **98**(3), 457 – 470.

Wilson, V. J., and Melvill Jones, G. (1979). *Mammalian Vestibular Physiology*. New York: Plenum.

Witmer, B. G., and Singer, M. J. (1998). Measuring presence in virtual environments: A presence questionnaire. *Presence: Teleoperators & Virtual Environments*, **7**(3), 225–240.

Wray, M., and Hawkes, R. (1998). Distributed virtual environments and VRML: an event-based architecture. *Computer Networks and ISDN Systems*, **30**(1-7), 43 – 51.

Yardley, L., and Higgins, M. (1998). Spatial updating during rotation: The role of vestibular information and mental activity. *J. Vestib. Res.-Equilib. Orientat.*, **8**(6), 435 – 442.

Glossary

This glossary should help readers from other fields (biology, computer science or psychology) to understand the specific meaning of terms in this thesis. It is a list and explanation of special words (e. g., technical, obsolete). Some of the definitions given are based on Hornby (1983).

acceleration In the mathematical sense, the time derivative of ↑velocity and the time integral of ↑jerk. It is proportional to the force applied to a free constant mass.

ACE Application Communication Environment is a library which allows one to program for multiple ↑OS's without knowing the details of different system calls. ACE wraps these differences effectively in providing a single programming interface.

actuator Term taken from the field of robotics to name a part of a machine which is actually causing changes in position. It is often used as a name for motors in a robot.

AD/DA Analog Digital/Digital Analog. To connect devices which provide analog currency to a digital computer and reverse, one needs to translate the specific current into a corresponding digital number. This process is called AD/DA conversion, respectively.

algorithm A finite set of rules that describes the transformation process of certain input variables into an output. Further, this description guarantees efficiency, termination and determination of the process (see Klaeren (1991)).

AGP Accelerated Graphics Port. A bus interconnect mechanism designed to improve performance of 3D graphics applications. AGP is a dedicated bus from the graphics subsystem to the core-logic chipset. It recently replaced the ↑PCI graphics cards interface due to the higher transfer rate.

architecture The classical meaning outlines the art of designing buildings. The computer scientist often refers to the architecture of a system as the general structure based on the single specific units which are necessary to build the overall system.

avatar In Hindu myths a deity in human or animal form descended on earth. In ↑VR it refers to a simulated person controlled by the human user displaying the user's entity in VR.

BNC Bayonet Norm Connector (also Bayonet Neill-Concelman). The thinnet or thin Ethernet cabling (RG-58 coaxial cable) with the BNC (metal push and turn-to-lock) connectors is technically called 10Base2.

canals Parts of the vestibular system. They primarily are sensitive to angular accelerations. In combination with ↑otoliths they form part of the somatosensoric body sense of animals and humans.

client Normally the customer of professionals and similarly used in computer science to name the program which connects to a ↑server in order to get or send data.

coefficient In the mathematical sense, the number or symbol placed before and multiplying another quantity.

communication The act of passing on information. It refers in this thesis to the process which transfers data from one computational unit to another.

correlation Expresses the mutual relationship between quantities. See appendix A for a mathematical definition.

CPU Short for the Central Processing Unit of a computer. It refers to the chip which actually does the computation specified by the program.

CRT Short for cathode ray tube. The electron tube is used for the visualization of fast electrical oscillations, as K. F. Braun expressed it in 1897. It is used in oscilloscopes as well as in modern TV screens and computer monitors.

cue A hint on how to behave and what to do. In psychophysics, it often refers to one specific feature of the sensation of a modality.

CVS Center for Visual Science in Rochester, NY or Concurrent Versions System. A database system which allows the reconstruction of previous versions of a file based on the storage of differences.

DACS Distributed Application's Communication System. It provides different communication primitives like ↑RPC, ↑message parsing and data streams. For details see Jungclaus (1998).

dB Decibel. Unit to measure the relative loudness of sound on a logarithmic scale.

device Something used for a special purpose. In computer science it is used for hardware parts.

display Name of device used for showing the internal state of a computer.

distance Measure of space between two points or places.

distributed Being put in different places; computers that are connected via a network and share common resources. Distributed systems share resources, data or work load.

DOC++ A tool for documentation done in the source code of Java or C/C++. It is capable of extracting class, data and function declarations in combination with specialized comments. The program generates ↑HTML for online access or LaTeX output for printing.

DOF Degree Of Freedom is the mathematical expression for the number of possible independent axes (dimensions) of a vector space.

DSP Short for Digital Signal Processing. Fast chips implement digital filters and other functionality in hardware for high data throughput.

EEG ElectroEncephaloGram; non-invasive method to measure electrical potentials from cortex.

Ethernet Name of a computer network standard. It was once believed that Ether was the medium though which light waves were transmitted through space.

force Strength or physical power. It is proportional to the ↑acceleration of a constant mass and could therefore be used by the subjects in chapter 4.

frequency Number of occurrence of an event per time interval. The unit Hertz [Hz] specifies the amount per second.

gain-factor Specific coefficient for one modality in psychophysical experiments.

Gauß Karl Friedrich Gauß was mathematician and astronomer (1777-1855). He developed a probability theory of observation errors. The Gaussian distribution is named after him.

GNU GNU's Not Unix. See GNU[2]

graph A diagram displaying the variation of two or more quantities.

graphic Visual symbolisation of letters, diagrams or drawings. In computer science the displayed information in form of pictures on a computer monitor is called computer graphics.

hardware Parts of computer equipment that are, anecdotally, damaged when dropped on the floor.

heading The direction of travel in world coordinate system as compared to ↑yaw which is the same value specified in the persons coordinate system.

heave Specifies the amount of raise or lift up (up and down movements of the ↑platform). It is used in combination with ↑surge and ↑sway, describing linear movements.

HMD Short for Head Mounted Display. A helmet which enables the user to see a computer generated picture in mono or stereo.

HTML Short for Hyper Text Markup Language, a computer language for description of layout and design of documents on the ↑WWW.

I/O Defines the input and output mostly in connection to a computer.

information Entity of knowledge, the contents of a message either in syntactical or semantic meaning.

IPD The inter pupilar distance. It is adjustable in most ↑HMD's in order to present the simulated picture with the appropriate distance between the eyes.

IRIX UNIX-like ↑OS from ↑SGI running on SPARC ↑CPU based computers.

ISA Short for Industry System Architecture. Old technology which was replaced with ↑ PCI.

jerk In the mathematical sense, the time derivative of ↑acceleration.

jitter Random noise often applied to distance variables in order to hide the true location.

joystick Computer interface which allows control of analog positions in multiple axes. See section 2.2.4 for examples.

judgement Used here as an estimation of magnitude.

kinematics Part of mechanics which describes movements without taking the necessary forces into account. In robotics, the kinematics is the definition of possible movements of a robot arm normally defined by linear matrix operations describing linear or angular changes of joints.

LAN A Local Area Network is limited to small areas mostly being a building or campus. It is defined in contrast to a ↑WAN.

---

[2]http://www.gnu.org/

latency Time difference between the cause and the occurrence of an event. In reality, latencies can be observed, for example, as an effect of speed of transmission.

LCD Short for Liquid Crystal Display. For each pixel of a computer generated picture, this display technology connects one black or three colored elements which transparently shield a white background illumination. It recently became an alternative for ↑CRT based computer monitors.

library Used for the collection of programming functions provided together with a declaration of the functions in so called header files.

LIFO The "last in first out" principle is commonly implemented in ↑stacks. The last value stored is the one first provided for output. It is the opposite of FIFO (first in first out) which is the principle for queues.

Linux Kernel and basic part of a free UNIX-like ↑OS which provides, in combination with the ↑GNU tools, a complete ↑OS. It is developed across the Internet by thousands of interested computer scientists and hobby programmers. The start was made by Linus Torvalds in 1991 by providing the basic concepts and implementation for the kernel.

MCC The Motion Control Card (MCC) is part of the Maxcue motion platform system. It is a ↑DSP board which implements parametrized digital filters for motion cueing algorithms and simple smoothing of the movements of the motion platform.

MDU The Motion Drive Unit (MDU) is part of the Maxcue motion platform system. It amplifies the provided signals for the motors by the ↑MCC which controls the leg lengths of the motion platform.

message parsing Communication method which transfers the content together with its declaration e.q. its semantic definition.

modality The sensor organs of the human body are modality specific in the sense that they are sensitive for one specific physical property of the perceived stimulus.

MPI Max-Planck-Institute ;-)

MTBF The "mean time between failures" specifies the expected time difference between failures to specify the robustness to failures of a ↑device.

NIH National Institute of Health. American scientific funding agency. Part of the US Government.

noise (Sound) signal which is uncorrelated to time. So called "white noise" has a mean of zero.

NTSC National Television Standards Committee. An American video format norm.

observer Someone who sees and notices or watches carefully. In the context of this thesis, the observer refers either to the human subject of an experiment perceiving the simulation or to the virtual observer which is the symbolic entity of the human observer in the simulation program.

Onyx High-end computer build by ↑SGI. See section 1.1.3 for examples.

OpenGL Complex 3D-Graphics Library initially designed by ↑SGI. There is special hardware available which handles the commands from the graphics rendering stack of OpenGL. Software emulations like MesaGL enable a user without special hardware to use the interface for 3D graphics.

OpenGVS Graphics library from Quantum3D extending ↑OpenGL with more complex features. The library is available for WindowsNT and ↑Linux. A rendering tree enables the setup of complex, dynamic scenes. See ↑Performer and ↑Vega as comparable solutions.

otoliths Part of the vestibular system in the inner ear. The otoliths are small crystals embedded in gelatinous mass around sensor cells. They are influenced by linear acceleration, due to the density difference to the surrounding medium. See ↑canals for the other part of the vestibular system.

OS The operating system of a computer allows the user to run programs and to manage all connected resources and devices.

paradigm A) Vendor of ↑Vega; B) Experimental philosophy guiding the experimental design.

PC Short for Personal Computer in opposition to large mainframe computers which were common before the invention of the PC.

PCI Peripheral Component Interconnect. 32-bit bus designed by Intel to be the successor of ↑ISA. This bus allows the extension of a ↑PC by additional equipment with high speed access.

perception Process and result of the sensation by which we become aware of changes through the senses of sight, hearing, etc..

performance Noticeable action or achievement. In computer science, a speed rating is used to measure performance on a objective scale.

Performer Graphics library from ↑SGI extending ↑OpenGL with more complex features. The library is available for Linux and ↑IRIX. A rendering tree enables the setup of complex, dynamic scenes. It also builds the base for ↑Vega. See ↑OpenGVS as comparable solution.

pitch Rotation around the axis connecting one's ears. See also ↑roll and ↑yaw.

platform Refers to the upper movable part of the motion platform in contrast to the fixed base part.

plot Same as ↑graph or diagram.

proprioception The combination of the perception of muscle spindles and joint flexion receptors of the body. In combination with the vestibular system, it constitutes the somatosensoric sense.

psychophysics Description, quantification, and interpretation of perception as defined by G. T. Fechner in 1860.

PVM The parallel virtual machine enables the distribution of work load across a heterogenous network of computers for parallel execution.

RAM Random access memory; refers to the working memory of a computer.

roll Turn around the axis between the tip of the nose and the back of the head. See also ↑ pitch and ↑yaw

RPC Short for remote procedure call, which allows the start of processes on a distant computer. The technique is often base for high level distributed mechanisms like distributed shared memory.

RS232 A very old serial line protocol standard (same as EIA-232).

scheduling The method of distributing work capacity among processes running on the ↑ CPU(s) of a computer.

script Short program in an interpreter (sometimes shell) language. In comparison to compiled programs, scripts are much slower, but more flexible due to the fact that they are interpreted at runtime.

server A computer with a special program which provides a service used by different ↑ clients. Sometimes the program itself is referred to as a server.

SGI Silicon Graphics Inc., vendor of ↑IRIX and ↑Performer.

software Part of a computer system which exists as the logical description of data or programs. It is necessary to have ↑hardware in order to use software.

spatial In relation to or existing in space. Spatial ↑cues form a contextual representation of the space around us, which is called spatial frame of reference.

stack Computer science mechanism of storing data in a ↑LIFO fashion. It is used, for example, for recursive function calls. The program's return address is saved on a stack in order to jump back at the end of a subroutine.

Stewart British engineer who proposed in 1965 "a platform with six degrees of freedom." (Stewart, 1965).

subject The typical patient human participant in my experiment.

surge The linear forward and backward movements of the ↑platform. See also ↑heave and ↑sway.

sway The lateral, left and right movements of the ↑platform. See also ↑heave and ↑surge.

task Either a process (meaning autonomous program) on a UNIX running computer or the ↑subjects' assignment in an experiment.

TCP transmission control protocol for the Internet which allows the confirmation of a data package through handshake mechanisms. This is in contrast to ↑UDP, which does not guarantee the arrival of a single package.

TCP/IP The Internet Protocol based on ↑TCP.

TFT Thin Film Transistor. A special form of ↑LCD with high luminance and strong colors. TFT displays can be seen from a large viewing angle.

thread Part of a parallel program. Threads are implemented either on ↑OS level or library level using separated processes to emulate thread functionality. Threads of a program can access shared memory and allow parallel execution on multiprocessor systems.

UDP User Datagram Protocol. Unsafe protocol for network communication. See also ↑ TCP.

VE Short for Virtual Environment. The overall simulated context in a ↑VR application.

Vega Graphics library from ↑Paradigm extending ↑Performer with more complex features. The library is available for WindowsNT and ↑IRIX. A rendering tree enables the setup of complex, dynamic scenes. A comfortable user interface allows the development of ↑VE by click and drop. See ↑Performer and ↑OpenGVS as comparable library solutions.

velocity In mathematical sense the time derivative of ↑distance and time integral of ↑acceleration. The human vestibular system probably codes linear and angular velocity although the sensor organs are sensitive to accelerations.

vestibular Concerning the inner ear organ that senses linear and angular accelerations.

virtual The Latin etymological root means existing by power and possibility, able to work or cause, apparently or seemingly. In combination with reality it outlines the simulation of multiple modalities for an observer.

visual Being perceived with the eyes.

VR Virtual Reality is "*... a high-end user interface that involves real-time simulation and interactions through multiple sensorial channels. These sensorial modalities are visual, auditory, tactile, smell, taste, etc.*" (Burdea, 1993). See also ↑VE.

WAN A Wide Area Network is defined in contrast to ↑LAN, meaning a bigger network, generally the size of a town or region.

whiskers Symbolize the long hairs growing near the mouths of rats or cats. They depict one standard deviation in some of the ↑plots.

WWW World Wide Web; the biggest distributed system of today. The most commonly used language for describing documents in the WWW is ↑HTML.

X11 The X Window System is a large and powerful graphics environment for UNIX systems. The original X Window System code was developed at MIT; commercial vendors have since made X the industry standard for UNIX platforms. Every modern UNIX workstation runs some variant of the X Window system.

XGA Same as XVGA and short for eXtended Video Graphics Adapter and the format of 1024x768 pixels.

yaw Turn around the body vertical axis. See also ↑pitch and ↑roll.

Appendix A

Statistics and math

This appendix should summarize the mathematical methods that were used for the statistical analysis. Most of the statistical tests were taken from Köhler, Schachtel, and Voleske (1995) and were considered to be applicable based on the given requirements. The tests themselves were calculated with the statistical package of `matlab` and the Unix ANOVA. Additionally, all of the experimental design of the stimulus conditions was done with `matlab` scripts. Furthermore, all the graphs were plotted using `matlab`.

The analysis of the data often used general correlations, which indicate how well data can be fitted with a linear function. The correlation coefficient r lead to a measurement of determination $B = \sqrt{r}$ which indicates how much of the variance in the data is explained by the linear fit. On the other hand, one can use a regression analysis, which assumes a unidirectional dependency of Y on X: $X \rightarrow Y$, in contrast to the correlation which does require any functional dependency. For the regression analysis the slope of the fit can be tested with a t-test against zero.

Another method which was used is the ANOVA (=ANalysis Of VAriance). The general idea of the ANOVA is to calculate levels of significance for the general hypothesis that the variation of single or multiple factors had an effect on the data. The levels of significance is the probability of obtaining the data given a (null-)hypothesis $(P(D|H))$. It does not specify the strength of an effect, the reproducibility of the data or the probability of any hypothesis (Haller, 1999). It is often assumed that the probability of the hypothesis and data are known. In this case, by application of the Bayes-theorem $(P(H|D) = \frac{P(D|H)*P(H)}{P(D)})$ one could calculate the probability of the hypothesis given the measured data.

Appendix B

More experimental data plots

This appendix illustrates more data obtained in the experiments. The large number of plots would often interrupt the flow of arguments if included into the main text. The following table will help to find the right plot[1].

Exp.	Plot Name	Figure
DVA	Perfect performance assumed for the judgments	B.1
	Pooled data for all subjects plotting judgments against physical values. All data for X = forward-backward linear movements.	B.2
	Same, but all data for Y = left-right linear movements.	B.3
	Same, but all data for H = left-right turns (heading).	B.4
	Model fit for the pooled data for all subjects plotting judgments as function of both varied factors. All data for X = forward-backward linear movements.	B.5
	Same, but all data for Y = left-right linear movements.	B.6
	Same, but all data for H = left-right turns (heading).	B.7

Table B.1: More data plots for the experiments: DVA = Distance, Velocity and Acceleration judgments.

[1]In the PDF version one can follow the links...

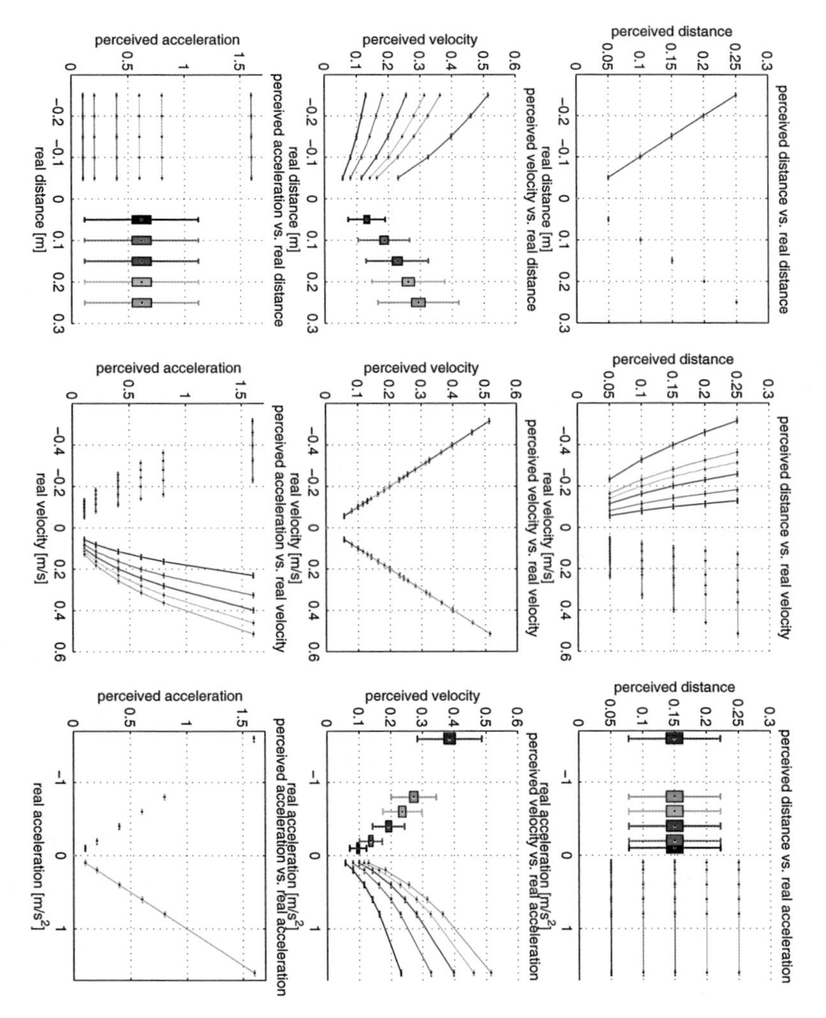

Figure B.1: Perfect Performance which is calculated based on the physical stimulus.

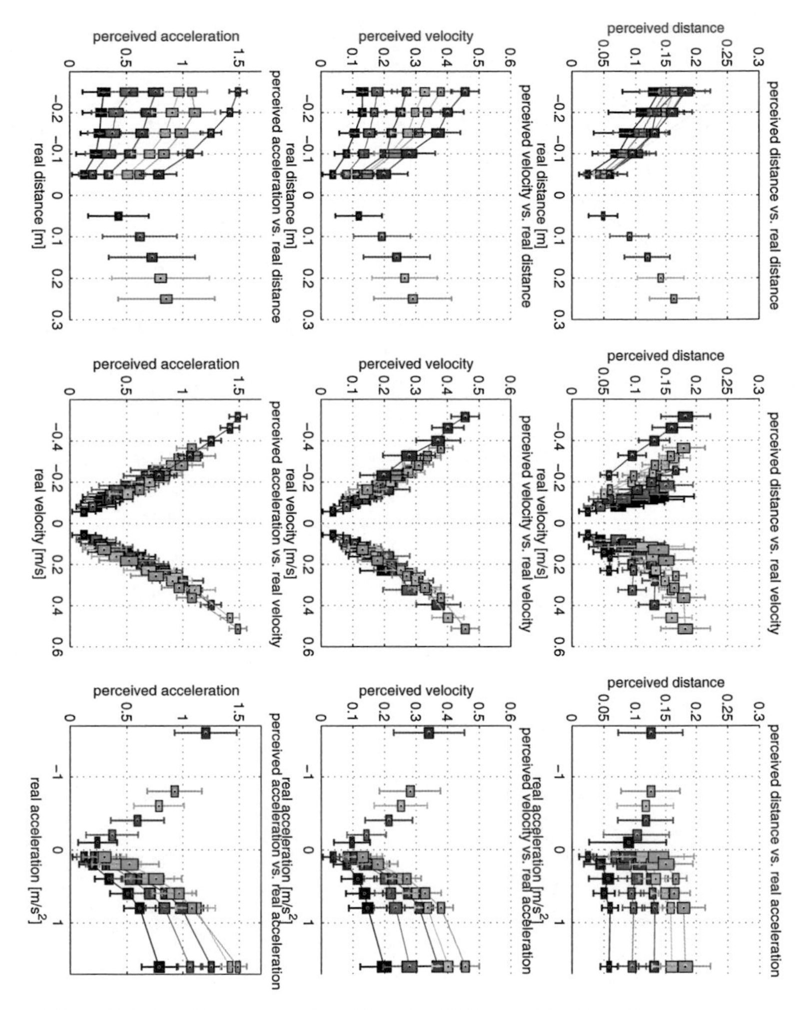

Figure B.2: Pooling accross all subjects for X (forward-backward-movements).

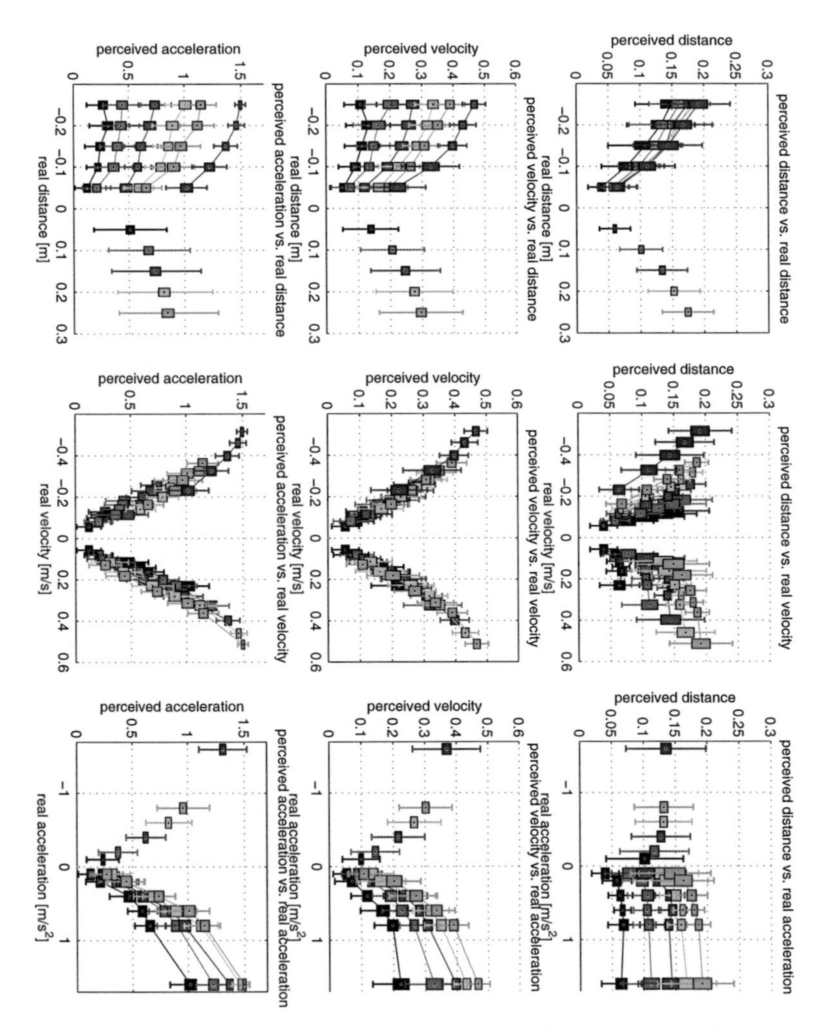

Figure B.3: Pooling accross all subjects for Y (left-right-movements).

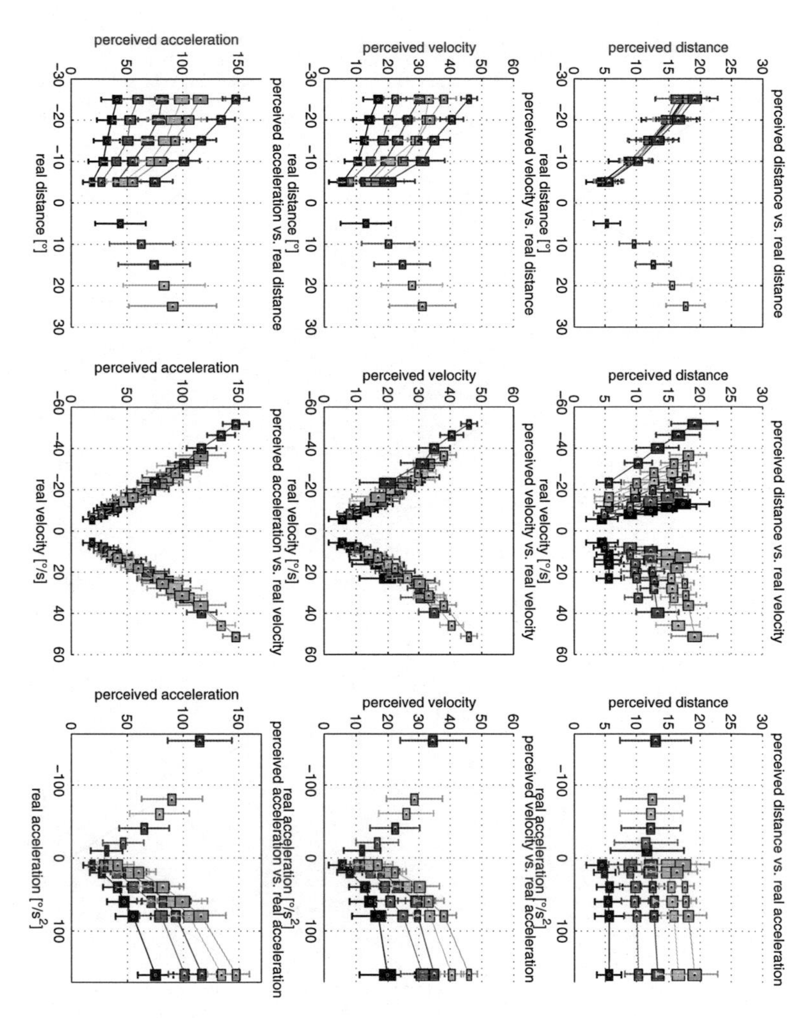

Figure B.4: Pooling accross all subjects for H (turns around the body axes).

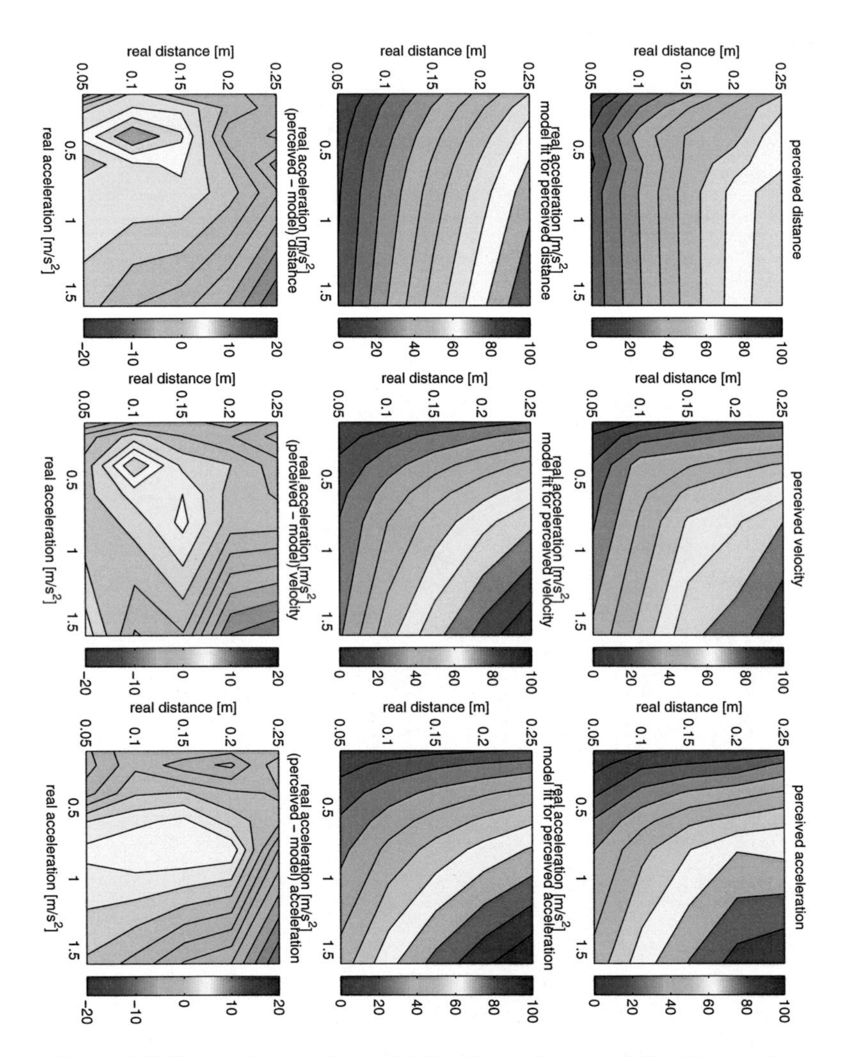

Figure B.5: The graphs show the model fit to the pooled data of all subjects to the X (forward-backward) direction (see Fig. B.2). Each plot in the first row displays one block (distance, velocity, and acceleration judgements) of the experiment. The color encodes the mean of the subjects' judgements across all factorial combinations. The second row is the result of the fitted model to the data of the first row. The last row depicts the difference between the model and the data.

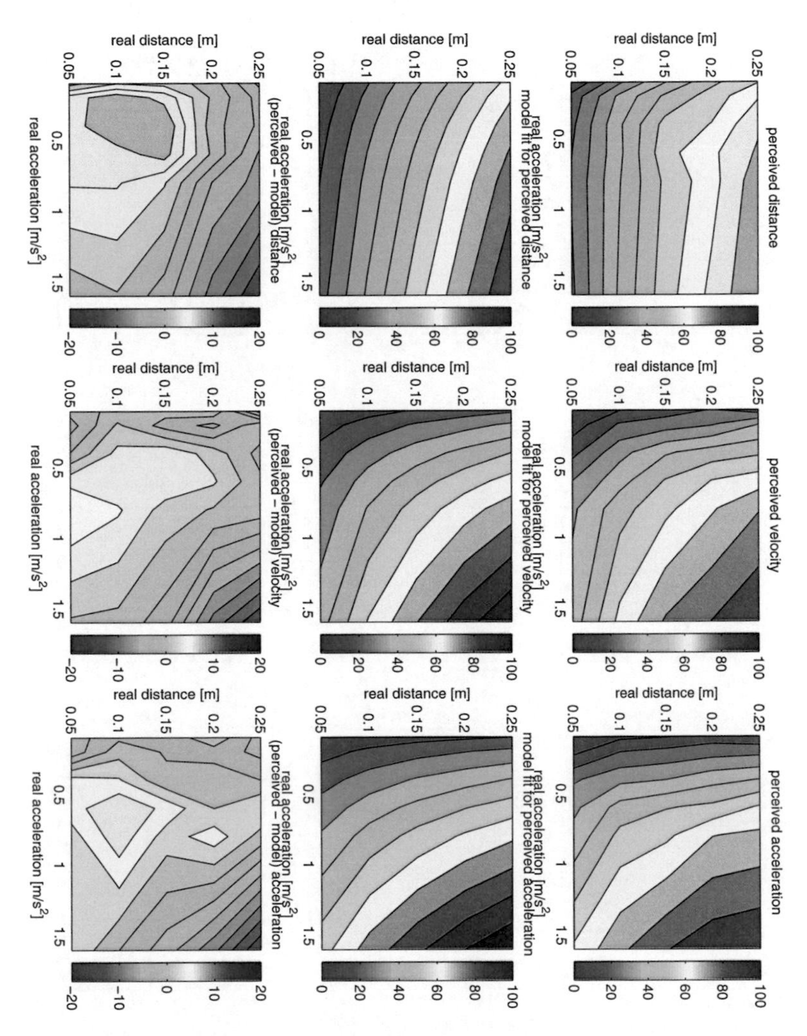

Figure B.6: The graphs show the model fit to the pooled data of all subjects to the y (left-right) direction (see Fig. B.3).

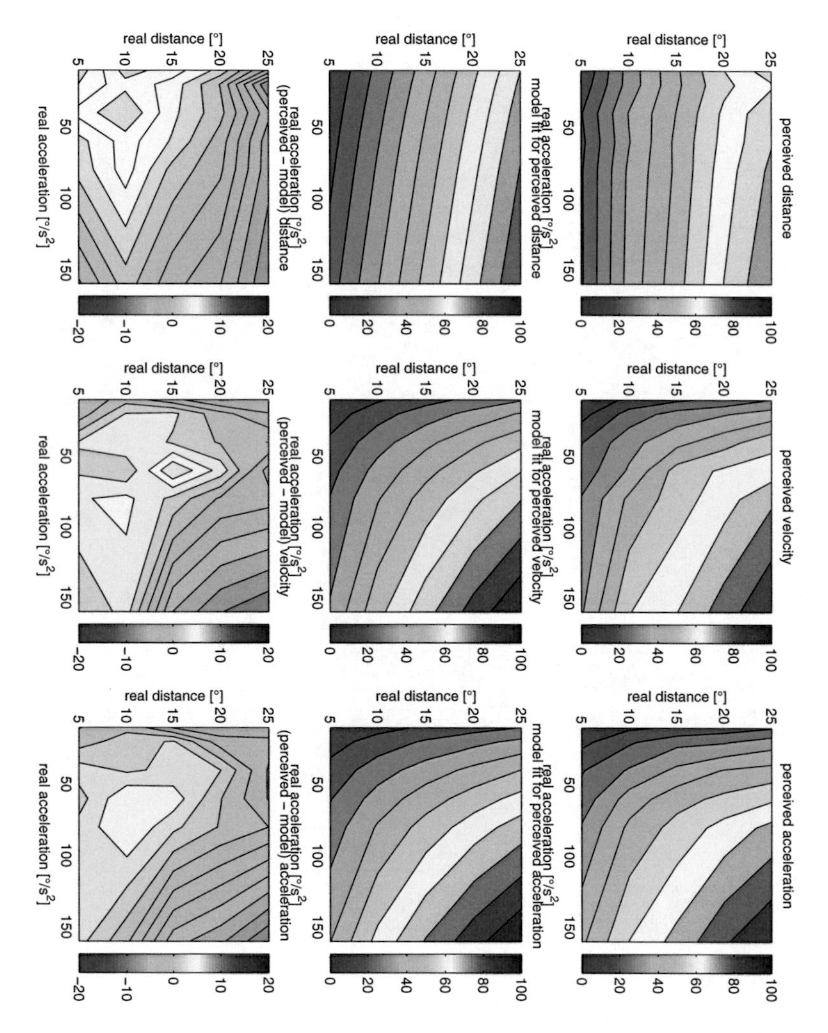

Figure B.7: The graphs show the model fit to the pooled data of all subjects to the H (turns around the bogy axes) movements (see Fig. B.4).

Appendix C

Technical Data

The technical data given in this appendix are based on the manufacturers data. Due to frequent changes on the technical data for products available, the data for the actual hardware equipment are printed here. However, there is no guarantee that the actual performance matches this description of the systems given by the manufacturers. The following table C.1 points towards the specific table describing the technical data.

Equipement	**Product Name**	**Manufacturer**	**Table**
motion platform	Maxcue	Motionbase	C.2
head mounted display	Proview50XL	Kaiser Electro-Optics	C.3
tracker	IS600mk2	Intersense	C.4
analog I/O card	AT-MIO-16E-10	National Instruments	C.5
sound card	Sound Blaster *live!*	Creative Labs	C.6
headphone	Aviation Headset HMEC 300	Sennheiser	C.7
force transducer	VT-2	RHB	C.8
Computer Name	**Purpose**	**OS**	**Table**
Sprout	driving dynamic	IRIX 6.5	C.9
Cantaloupe	analog I/O	Linux	C.10
Cucumber	main simulation	Linux	C.11
Borage	platform control	Windows 95	C.12
Soy and Tofu	graphics	WindowsNT	C.13

Table C.1: Technical data overview

Feature	Specification
Payload	1000kg (2200lbs)
Degrees of freedom	6
Power requirements	200-250VAC single phase 50-60Hz 35A (peak current) 208VAC three phase 50-60Hz 25A (peak current) 110VAC at reduced peak speed (0.3m/s heave) Less than 2A when at rest in an elevated position.
Height when parked	1080mm (43in)
Base frame dimensions	1800mm x 1800mm (71in x 71in)
Lubrication	Sealed unit lubricated for life.
Actuator technology	Maintenance-free brushless motors, direct drive, zero backlash. 20,000 hour design life.
Actuator stroke	450mm (18in)
Actuator bandwidth	25Hz
Actuator position resolution	0.6μm (0.00003in)
Peak actuator thrust	13kN (2,900 lbf)
Surge range	930mm (37 in)
Sway range	860 mm (34 in)
Heave range	500 mm (20 in)
Pitch range	+34/-32°
Roll range	+/-28°
Yaw range	+/-44°
Peak heave velocity	+/-0.6m/s (+/-24in/s)
Peak surge/sway velocity	+/-0.7m/s (+/-28in/s)
Peak surge/sway/heave acceleration	+/-0.6g over whole motion envelope. Heave acceleration better than +1g/-2g near centre of motion envelope.
Peak pitch/roll rate	40°/s
Peak yaw rate	60°/s
Smoothness	Better than 0.02g (simultaneous actuator reversal)
Sound level	Less than 60dBA
Power supply unit	250mm (10in) high, 605mm (24in) wide, 740mm (29in) long; floor or wall mounted. 7m (275in) cable to motion platform, other lengths optional.
System controller	PC ISA bus processor card, occupying two slots. 1.5m (60in) cable to power supply unit.
Control software	1. Direct control of each degree of freedom (with ride-film synchronisation). 2. Real-time motion cues from vehicle state information provided by host.
Transport delay	Maximum system latency less than 16ms.
Interface to host	1. PC ISA bus via dual port RAM. 2. Ethernet.
Ambient operating temp.	0 to 40°C (32 to 104°F).
Humidity	10-95% RH (non-condensing)

Table C.2: Technical data of the Maxcue motion platform form Motionbase

Feature	Specification
	Display
Type LCD	Full color, active matrix TFT, high speed polysilicon LCDs
Resolution/Eye	XGA Resolution (1024H x 768V) 2.34 arcmin/color group
Brightness	5-50 fL (adjustable)
Contrast	40:1
	Optical
Field of View	50°diagonal, 30°(V) x 40°(H)
Transmission	Non see-through
Optics	Color-corrected, aspheric refractive lens Independent optical paths for each eye
Eye Relief	Eyeglasses compatible
Exit Pupil	Non pupil forming
Overlap	100%
Stereo/Mono	Yes
Color Coordinates	Red: u' = 0.5099 v' = 0.5228 Green u' = 0.1033 v' = 0.5774 Blue u' = 0.1314 v' = 0.2250
	Mechanical
IPD	Independent left/right
IPD Range	55 - 75 mm
HMD Weight	35 ounces (980 g)
Headtracker	Accommodates magnetic and inertial tracker sensors
	Control Unit
Video Inputs	One or two XGA 1024 x 768, H&V - TTL, Analog 0.7 V P-P, 75 ohms, 60 Hz video inputs Autosense for stereoscopic or monoscopic operation
Horizontal Scan Rate	48.363 kHz (Internal and external sync)
Vertical Scan Rate	60.004 kHz
Genlocked Inputs	Independent phased locked loops for left and right eye
Cable Length	20 feet
Video Outputs	2 XGA video loops to display monitor
Controls	Audio adjust
Indicators (LED)	Power on/off, Video in
Connectors	XGA, 15 pin DA, female, (video in) 2 XGA, 15 pin DA, female, (video out for monitor) 2 BNC barrel connectors, RGB H&V, (video in) 2 sets RCA connectors, (stereo audio com pass through) 2
Power	85 - 264 VAC, 47-440 Hz, 25W (power cable included)

Table C.3: Technical data of the Proview XL50 Head Mounted Display from Kaiser Electro-Optics

Feature	Specification
Fusion Mode Specifications	
Degrees of Freedom	6 (per station)
Resolution:	
Position (X/Y/Z)	2.5 mm RMS
Angular (P/R/Y)	0.10° RMS
Stability:	
Position (X/Y/Z)	7.0 mm RMS
Angular (P/R, Y)	0.25°, 0.5° RMS
Max update rate	Serial 115.2 k baud
1 station	180 Hz
2 stations	120 Hz
3 stations	90 Hz
4 stations	60 Hz
Genlock options	NTSC, TTL, internal sync
Prediction range	50 ms
Latency	4 - 10 ms (w/o prediction)
Interface	RS-232 up to 115.2 kbaud
Protocol	Industry standard protocols Compatible with IS-900/ IS-300
Tracking Coverage Area	3.5 m x 3.5 m
Physical	
Power	100-240 VAC, 60 W
Fusing	100-120 VAC: T250V, 1.0A 220-240 VAC: T250V, 0.5A
Operating Temperature	0 to 50°C (32 to 122°F)
Storage Temperature	-20 to 70°C (-4 to 158°F)

Device	Dimensions	Weight	Cable Length
InertiaCubeTM Sensor	26.9 mm x 34.0 mm x 30.5 mm	60.0 g	9 m
SoniDiscTM Position Sensor	25.4 mm x 25.4 mm x 16.5 mm	11.3 g	9 m
Long X-Bar Installed	1.42 m x 1.42 m x 0.04 m	3.7 kg	10 m
ReceiverPod (each)	0.12 m x 0.08 m x 0.04 m	0.36 kg	0.6 m
Base Unit Signal Processor	42.5 cm x 30.5 cm x 10.2 cm	3.81 kg	

Table C.4: Technical data of the IS-600 Mark 2 tracking system from InterSense

Feature	Specification
Analog Inputs	16 single-ended, 8 differential channels 100 kS/s sampling rate 100 kS/s stream-to-disk rate 12-bit resolution
Analog Output	2 channels, 12-bit resolution
Digital I/O	8 (5 V/TTL)
Counter/Timers	2 up/down, 24-bit resolution, digital triggering
Driver Software	NI-DAQ, Windows 2000/NT/9x
Application Software	LabVIEW, LabWindows/CVI, ComponentWorks, VirtualBench, Measure, BridgeVIEW, Lookout

Table C.5: Technical data of the AT-MIO-16E-10 analog card from National Instruments

Feature	Specification
Hardware	
Frequency Response	10Hz - 44KHz
Signal to Noise Ratio	>96 db
Noise Floor	-120dB
Sampling Rate for Playback/Recording (Stereo)	8 KHz - 48 KHz
Supply Voltage Requirement (Loading)	+5, +12, -12 Volt
Current Consumption (Typical)	300, 500, 30 mA respectively
Microphone Impedance	600 Ohms
Line-In Impedance	47 KOhms
CD Audio-In Impedance	50 KOhms
Microphone Sensitivity	10 - 200 mVpp
Line-In Sensitivity	0 - 2 Vpp
CD Audio-In Sensitivity	0 - 2 Vpp
AD/DA Resolution	16 bits
Environmental	
Environment Temperature (non-operating)	-40°C to 70°C
Environment Temperature (operating)	10°C to 50°C
Relative Humidity (non-operating)	30% to 95%
Relative Humidity (operating)	30% to 80%
MTBF	>60,000 hours
Drop Test	30cm above concrete ground on all 6 sides

Table C.6: Technical data of the Sound Blaster Live! sound card from Creative Labs.

Feature	Specification
Headphones	
Transducer	dynamic, closed, circumaural
Frequency response	45 - 15,000 Hz
Impedance active/passive	160/150 ohms, mono; 320/300 ohms per system, stereo
Attenuation (active and passive)	> 25 - 40 dB
Max. SPL	118 dB linear
Caliper pressure	approx. 10 N
Microphone with Preamplifier	
Type	MKE 45-1
Transducer	electret, noise-compensated
Frequency response	300 - 5,000 Hz
Max. SPL	120 dB (distortion < 5 %)
Output voltage	400 mV +/-3 dB at 114 dB/SPL (as per RTCA/DO-214)
Min. terminating impedance	150 ohms
Operating voltage	typ. 16 V DC (8-16 V DC, 8-25 mA) as per RTCA/DO-214
General Data	
Part number	300-231-415
Connection cable	single-sided round cable, length 1.5 m
Weight without cable	370 g
NoiseGard[tm] supply	12 - 35 VDC
Current consumption	approx. 27 mA, max. 80 mA
Fuse	500 mA multifuse
Connectors	headphones: 1/4” stereo jack, microphone: PJ-068, XLR-3 for NoiseGard[tm] supply
Specials	mono/stereo switch, on/off switch for NoiseGard[tm], headphone volume control
Operating temperature	-15 to +55°Celsius
Storage temperature	-55 to +55°Celsius

Table C.7: Technical data of the Aviation Headset HMEC 300 from Sennheiser

Feature	Specification
	FX-80 Motion Actuator
Frequency Response	15Hz-90Hz
Resonant Frequency	43 Hz
Recommended Power	20-150 Watts
Impedance	4 Ohm
Force	60 Lbf @ 100 Watts
Dimensions	$4\frac{3}{4}$ Dia. x $2\frac{3}{4}$" H
Weight	3 Lbs.
	SAM-200 Amplifier
Amplifier Type	Discrete Class A/B
Power Output	200 Watts continuous into 4 ohms @ 1% THD
Distortion(THD)	$<$ 0.15% @ 1 Watt
S/N Ratio	$>$75 dB (without filter)
Input Impedance	45 Kohms (line level input), 200 ohms (speaker level input)
Input Sensitivity	18 mV (line level input), 100 mV (speaker level input)
Crossover Slope	12 dB/octave
Crossover Freq. Range	40Hz-180Hz (-3dB)
Auto Turn-on Sensitivity	6mV @ 50Hz
Turn Off Delay	15 Minutes
Frequency Response	10Hz-40kHz (-3dB) (crossover off), 10Hz-Variable from 40Hz-180Hz (-3dB) (crossover on)
Damping Factor	$>$100
Dimensions	17" W x 4" H x 13" D
Weight	17 Lbs.

Table C.8: Technical data of the Virtual Theater Kit VT-2 from RHB. The VT-2 consist of one SAM-200 Amplifier and two FX-80 Tactile Transducers.

Feature	Specification
CPU	250 MHZ MIPS R4400
Memory	256 MB
Graphics	High Impact
SCSI	WD33C93B
Sound	Iris Audio Processor
Operating Systems	IRIX 6.5

Table C.9: Sprout: Indigo^2 from SGI

Feature	Specification
CPU	450 MHz Pentium III
Memory	128 MB
Graphics	Texas Instruments TVP4020 Permedia 2
Motherboard	Intel 440BX
SCSI	Adaptec AIC-7890/1 Ultra2 SCSI host adapter
CDROM	Toshiba XM-6401TA
Harddisc	13 GB, IBM DDRS-39130D
Network	Intel Speedo3 Ethernet
Sound	Soundblaster *live!* EMU10K1
Operating Systems	Debian 2.2 GNU Linux

Table C.10: Cantaloupe: Linux Wheel PC

Feature	Specification
CPU	2*450 MHz Pentium III
Memory	256MB
Graphics	Diamond Viper 770 - TNT2 chip
Motherboard	ASUS P2B-DS (Intel 440BX)
SCSI	Adaptec AIC-7890/1 Ultra2 SCSI host adapter
CDROM	Toshiba XM-6401TA
Harddisc	9 GB, IBM DNES-309170W
Network	Intel EtherExpress Pro 10/100 Ethernet
Sound	2*Soundblaster *live!* EMU10K1
Operating Systems	Debian 2.2 GNU Linux

Table C.11: Cucumber: Linux Simulation PC

Feature	Specification
CPU	Celeron-MMX 433 MHz
Memory	128 MB
Graphics	Matrox Millenium G200
Motherboard	Intel440BX
SCSI	Adaptec AIC-7890/1 Ultra2 SCSI host adapter
CDROM	Teac CD-532S
Harddisc	13 GB, IBM DDRS-39130D
Network	Intel EtherExpress Pro 10/100 Ethernet
Operating Systems	Windows 95

Table C.12: Borage: Platfrom control PC

Feature	Specification
Industrial Chassis	18 Gauge Steel 4U industrial rack mount chassis; shock mounted drive bay; retention mechanism for CPUs and add-in boards; mount ears standard; 6.81 ins H x 21.50 ins D x 16.88 ins W
Forced Air Cooling	Dual 120mm, 150 CFM Bezel Fans with removable filter. Integral ball bearing power supply, CPU and graphics subsystem fans
Power Supplies	400W, 120/240 VAC 1 phase 50-60 Hz auto-sensing input
Motherboard	Intel GX+
CD	32X ATAPI EIDE CD-ROM
Floppy Drive	3.5" 1.44 MB Floppy (Black Bezel)
Operating Specifications	Operating shock: 2G (2ms @ 1/2 sine wave) Operating vibration: 0.25G (3.5-500 Hz sine sweep, 0 to peak) Operating temperature: +0°C +/-50°C with relative humidity 10-90%; non-condensing
System MTBF	25,000 hours per system (calculated)
Safety and EMI	FCC A, CE, and ETL certified
Peak Realtime 3D Graphics I/F Bandwidth	399 MB/sec (1 x 33 MHz, 32-bit PCI; 1 x 66 MHz, 32-bit PCI)
RT3D Graphics Subsystem	2 x Obsidian2 200SBi-8440 (3D Only)
2D/VGA Options	Integrated Cirrus Logic 2D/VGA with 4 MB SDRAM
System Memory	512 MB (ECC PC-100 DIM SDRAM)
CPU	2 x 500+ MHz Intel Pentium-III Slot-1 Processor(s)
Inter-channel Synchronization	SwapLock and SyncLock via custom cable assembly available with graphics systems.
Standard Disk Drive Options	8.4 GB Ultra Wide SCSI-II
LAN & WAN Options	Ethernet: Integrated PCI 10/100 NIC Standard
Operating Systems	Windows NT 4.0 SP6 (optional Linux)

Table C.13: Soy & Tofu: Heavy Metal GX+ Graphic Systems from Quantum3D

Appendix D

WWW adresses

The WWW links included in the thesis run the risk of being quickly outdated. In this case, one can look up the Motion-Lab Pages[1] and their given references which should be more up to date. Note that in the PDF version of this document, the given links allow direct access to the references.

D.1 Free software

MBROLA Multilingual Speech Synthesizer[2]

ACE: The Adaptive Communication Environment[3]

LinuxLabProject[4]

DOC++[5]

CVS[6]

GNU[7]

Debian[8]

Performer for Linux[9]

LaTeXon Dante[10]

comming soon: Motion-Lab Library

[1] http://www.kyb.tuebingen.mpg.de/bu/people/mvdh/motionlab/index.html
[2] http://tcts.fpms.ac.be/synthesis/
[3] http://www.cs.wustl.edu/~schmidt/ACE.html
[4] http://www.llp.fu-berlin.de/
[5] http://www.imaginator.com/doc++
[6] http://www.cvshome.org/
[7] http://www.gnu.org/
[8] http://www.debian.org/
[9] http://www.sgi.com/software/performer/linux.html
[10] http://www.dante.de/

D.2 Commercial software

IRIX Performer[11]

OpenGVS[12]

Multigen & Vega[13]

D.3 Hardware

Company	Product	is used for:
3Com[14]	SuperStack II Switch 3300[15]	the network connections
PolyCon[16]	PolyCon/S[17]	console switching hub
APC[18]	Smart-UPS 2200[19]	uninterruptible power supply
Quantum3D[20]	Heavy Metal[21]	visualization PCs with anti-aliasing
Motionbase[22]	Maxcue[23]	6-DOF motion platform
Kaiser Electro-Optics[24]	ProView XL50[25]	head mount display
Intersense[26]	IS-600mk2[27]	6-DOF tracking system
Microsoft - Hardware[28]	SideWinder[29]	force feedback joystick
National Instruments[30]	AT-MIO-16E-10[31]	steering force control
Creative[32]	SB Live![33]	spatial 3D sound
SENNHEISER[34]	Aviation Headset HMEC 300[35]	subjects headphones
RBH[36]	Virtual Theater® Kit VT-2[37]	force transducer for vibrations

[11] http://www.sgi.com/software/performer/
[12] http://www.opengvs.com/
[13] http://www.multigen.com/index.html
[14] http://www.3com.com/
[15] http://www.3com.com/products/dsheets/400260a.html#5
[16] http://www.polycon.com/
[17] http://www.polycon.com/content/produkte/polycon_s.html
[18] http://www.apcc.com/
[19] http://www.apcc.com/products/techspecs/index.cfm?base_sku=SU2200RMI3U
[20] http://www.quantum3d.com/
[21] http://www.quantum3d.com/product%20pages/heavy_metal.html
[22] http://www.motionbase.com/
[23] http://www.motionbase.com/html/page15.html
[24] http://www.keo.com/
[25] http://www.keo.com/ProView_XL50_spec_sheet.html
[26] http://www.isense.com/
[27] http://www.isense.com/products/prec/is600/index.htm
[28] http://www.microsoft.com/products/hardware/
[29] http://www.microsoft.com/products/hardware/sidewinder/devices/FFBpro/default.htm
[30] http://www.natinst.com/
[31] http://sine.ni.com/apps/we/nioc.vp?lang=US>pc=mn>cid=1067
[32] http://www.soundblaster.com/
[33] http://support.soundblaster.com/specs/audio/live/
[34] http://www.sennheiser.com/
[35] http://www.sennheiser.com/headsets/ac/ac_1d006.htm
[36] http://www.rbhsound.com/
[37] http://www.rbhsound.com/virtual/vt-2.htm

D.4 VR/VE-Labs

The labs mentioned in 1.1.3 do have a web-page describing some projects more in detail.

VE Lab – Tübingen[38]

VR Lab – Bielefeld[39]

NIH Resource Laboratory – Rochester[40]

Space Perception Lab – Santa Barbara[41]

Bankslab – Berkeley[42]

VRlab – Umeå[43]

D.5 Misc

Virtual Reality in Surgical Education[44]

AVANGO: A Distributed Virtual Reality Framework[45]

D.5.1 Tracking devices

Products from Polhemus[46]

Ascension[47]

Zebris[48]

Optotrak from Northern Digital Inc.[49]

[38]http://www.kyb.tuebingen.mpg.de/bu/projects/vrtueb/index.html
[39]http://www.techfak.uni-bielefeld.de/techfak/ags/wbski/labor.html
[40]http://www.cs.rochester.edu/u/bayliss/nih/resource.html
[41]http://www.psych.ucsb.edu/research/recveb/New_Pages/Areas_of_Research.htm
[42]http://john.berkeley.edu/
[43]http://www.vrlab.umu.se/
[44]http://www.vetl.uh.edu/surgery/vrse.html
[45]http://imk.gmd.de/docs/ww/ve/projects/proj1_2.mhtml
[46]http://www.polhemus.com/ourprod.htm
[47]http://www.ascension-tech.com/
[48]http://www.zebris.de/
[49]http://www.ndigital.com/optotrak.html